Enabling Asia to Stabilise the Climate

Shuzo Nishioka
Editor

Enabling Asia to Stabilise the Climate

Editor
Shuzo Nishioka
Institute for Global Environmental Strategies (IGES)
Kanagawa
Japan

ISBN 978-981-287-825-0 ISBN 978-981-287-826-7 (eBook)
DOI 10.1007/978-981-287-826-7

Library of Congress Control Number: 2015957815

Springer Singapore Heidelberg New York Dordrecht London

Printed on acid-free paper

Springer Science+Business Media Singapore Pte Ltd. is part of Springer Science+Business Media (www.springer.com)

Contents

Introduction: Enabling Asia to Stabilise the Climate

A Stable Climate Is a Common Asset for Humankind

The 5th Assessment Report (AR5) published by the Intergovernmental Panel on Climate Change (IPCC) in 2013 and 2014 revealed that temperatures will continue to rise as long as anthropogenic greenhouse gases (GHGs) are emitted into the atmosphere, and that the climate will not stabilise unless GHG emissions can ultimately be brought down to zero. AR5 also warned that we are in a critical situation, and if we continue to emit the current amounts of GHG, there is only 30-year quantity of GHG that can be emitted if we want to prevent a temperature rise less than 2 ° C from preindustrial levels.

A stable climate is a precious common asset for humankind. Local climates are incorporated into one comprehensive climate system at the earth's surface. Therefore, we cannot secure this common asset unless all countries take individual responsibility to deal with GHG emission reduction. Climate stabilisation is something that must be taken up by every country under the United Nations Framework Convention on Climate Change (UNFCCC).

A social transformation for climate stabilisation is the most significant worldwide challenge this century, and no country has experienced such a challenge before. We need to completely shift our social trend away from highly energy-dependent technology societies, a path that we have been on for 250 years since the Industrial Revolution, and turn our efforts in the direction of low-carbon societies within 50 years. Only then can we finally achieve zero-GHG-emission societies. There is not much time remaining to achieve this goal.

The Responsibilities and Role of Asia Are Vital

Asia has a very significant role and responsibilities for climate stabilisation. If Asia continues its current development in the form of highly energy-dependent societies, it is predicted that Asia will make up half the share of worldwide economic power, energy consumption, and carbon dioxide emissions in 2050. It is no longer possible for developing countries with their rapid economic growth to follow the path trod by developed countries to become highly energy-consuming technology societies. If the present infrastructure development and industrial investment follow the conventional pattern, developing countries in Asia will be locked in to a high-carbon-emission pathway for another half-century. Therefore, Asian countries themselves need to explore a path of development different from that followed by developed countries, and achieve leapfrogging to low-carbon societies.

An Opportunity to Leapfrog by Integrating Knowledge and Wisdom In-Country

The worldwide transition to low-carbon societies is a massive undertaking and it is up to each Asian country to set a vision for future society. Therefore, each country needs to form policies for national and local development by utilising its in-country knowledge without relying on others. It is necessary for each country to understand its specific situation and explore a future vision with the citizens who love their own country, with their full ownership.

This is a historical challenge that Asia is facing and, at the same time, it is a perfect opportunity for Asian countries to lead the development of a low-carbon world.

Good Practices of Science-Based Climate Policy Development Making Progress in Asia

Under these circumstances, this book aims to outline the challenges faced by each Asian country on how they are progressing in building up low-carbon societies, and it aims to share the information with other countries in the region and the rest of the world. By doing so, global cooperation for developing low-carbon societies can be further promoted.

The first part of this book clarifies that Asia holds the key to worldwide climate stabilisation, and examines model analyses of China, India, Japan, Vietnam, and Asia as a whole, showing that there is large scope for achieving development while reducing GHG emissions.

The second part introduces good practices showing how results of the examination of model analyses are actually incorporated into national and local-level low-carbon development policies and how they effectively work for policy formulation. For example, in Thailand, results from the model analysis have supported the Thai Intended Nationally Determined Contribution (INDC) to be submitted to the Conference of the Parties under the UNFCCC. This is a good example of national-level science-based policy formulation. On the other hand, in Malaysia, results from the model analysis on Iskandar were applied as a scenario in the development of a low-carbon society in Iskandar. Urban population is expected to account for 70–80 % of the worldwide population within this century. Therefore, it is likely that urban areas will take a front-line role in the formulation of low-carbon societies. The example of Iskandar shows one good practice in low-carbon society formulation.

The third part explains how to overcome barriers to measures implemented in each country's major policy sectors so that possible GHG emission reduction is actually realised by utilising good practices developed so far. Key categories for promoting decarbonisation are the promotion of public transportation, formulation of compact and energy-efficient cities, and forest conservation for enhancing carbon sinks and biomass energy use. Moreover, education and research communities are essential for formulating science-based policies. In this part, we present some advanced examples of how Asian countries are facing up to the challenges of leapfrogging to low-carbon societies.

International Cooperation for Knowledge-Sharing Towards Realising a Low-Carbon Asia

This book was written by experts and researchers who are making serious efforts to realise low-carbon development in Asia. On the way to low-carbon development is a very tough challenge that has never before been accomplished. Moreover, we have to lay out a new development pathway in a short span of time and then overcome various actual difficulties. Indeed, it will be a major contribution to the world if Asia can head in the direction of low carbon. However, there are three major obstacles to low-carbon development in Asia.

First, there is still no full-fledged system bridging science and policy to develop science-based policies. Second, policies responding to climate change have not yet been integrated into national development policies, and some policies have been formulated dependent on funds, resources, and knowledge from developed countries. Hence, in some countries, it is not sufficient to foster and make use of research communities in-country due to such constraints. Third, regional cooperation and collaboration are not fully matured as ways by which people can share similar environmental and developmental stages.

The only way to accomplish development that follows a new pathway in a very short time is to foster a research community in-country. In this way, we can promote science-based policymaking by facilitating discussions between policymakers and the research community, and go ahead with knowledge-sharing in the region by making full use of regional cooperation.

Japan has been conducting substantial international cooperation contributing to GHG reduction for approximately 20 years. However, it is high time for Asian countries to blaze a new trail towards realising low-carbon Asia under their own initiative. The Low Carbon Asia Research Network (LoCARNet) is a knowledge-sharing community, composed of researchers and those concerned who support the challenges being faced by developing countries themselves. With such network collaboration, Asia is making steady progress in the direction of low-carbon development. In 2014, at the LoCARNet annual meeting in Bogor, Indonesia, participants launched a declaration entitled "Asia is ready to stabilise the climate".

Utilise Asia's Full Force and Make the Leap to Stablise the Climate

The decade starting from COP 21 could well be the turning point for a major transition for world civilisation. In a business-as-usual scenario, Asia will account for about half of the world's economy, energy consumption, and CO_2 emissions in 2050, and if the region does not take this situation seriously, it will be impossible to make any significant global and historic changes. Now we are about to enter the age of substantial transition. For Asian countries that have not yet been locked in to high-carbon societies, it is indeed the very best opportunity to move forward to create a new low-carbon civilisation led by Asia.

In fact, this is what the world is very much looking forward to. In the past, it was Japan that succeeded in leapfrogging from the devastation of World War II to make a miraculous recovery and become an economic powerhouse. This was as the result of innovative technologies brought about by the oil crisis during the 1970s, which gave Japan a chance in a million to spring back from environmental problems associated with industrialisation and urbanisation and to overcome what had been an energy self-sufficiency rate of almost zero.

Asia can play a very significant role in turning the current climate crisis into opportunities for new development, so that the region can realise its potential and lead the way in low-carbon development. We will be more than happy if this book can provide confidence and hope to people not only in Asia but also across the whole world.

IGES
Kanagawa, Japan

Shuzo Nishioka

Part I
Asia Is a Key for a Sustainable Low-Carbon Society

Chapter 1
GHG Reduction Potential in Asia

Toshihiko Masui, Shuichi Ashina, Shinichiro Fujimori, and Mikiko Kainuma

Abstract Greenhouse gas (GHG) emissions from Asia accounted for approximately 38 % of global emissions in 2005. Considering the rapid economic growth expected in the coming decades, emissions from Asia in 2050 are projected to double the 2005 levels if efforts are not made toward achieving low-carbon societies (LCSs). The reduction of emissions in Asia is imperative for the transition by 2050 to an LCS worldwide that has halved GHG emissions. The LCS transition by Asian countries will not be an easy task. In order to accomplish this transition, it is vital that stakeholders including central and local governments, private sector enterprises, NGOs and NPOs, citizens, and the global community tackle it with a focused and common vision of the society they wish to achieve, while cooperating with one another and being aware of the roles they need to play. In addition, careful attentions should be placed on the diversity of the Asian countries when it comes to the implementation of countermeasures. Depending on the country or region in Asia, the level of development, amount and type of resources, climate conditions, culture, and other factors differ, and the actions that are effective may vary accordingly.

In order to analyze the feasibility, in this study two future scenarios, namely, advanced society scenario and conventional society scenario, are developed. In addition, "Ten Actions toward Low Carbon Asia," a guideline to plan and implement the strategies for an LCS in Asia, was developed. The ten actions are the following:

Action 1: Hierarchically connected compact cities
Action 2: Mainstreaming rail and water in interregional transport
Action 3: Smart ways to use materials that realize the full potential of resources
Action 4: Energy-saving spaces utilizing sunlight and wind
Action 5: Local production and local consumption of biomass
Action 6: Low-carbon energy system using local resources
Action 7: Low-emission agricultural technologies

T. Masui (✉) • S. Ashina • S. Fujimori
National Institute for Environmental Studies, Ibaraki, Japan
e-mail: masui@nies.go.jp

M. Kainuma
National Institute for Environmental Studies, Ibaraki, Japan

Institute for Global Environmental Strategies, Kanagawa, Japan

S. Nishioka (ed.), *Enabling Asia to Stabilise the Climate*,
DOI 10.1007/978-981-287-826-7_1

Action 8: Sustainable forestry management
Action 9: Technology and finance to facilitate achievement of LCS
Action 10: Transparent and fair governance that supports low-carbon Asia

The contributions of the ten actions have been quantified by a global computable general equilibrium model. The model outputs showed that GHG emissions in Asia can be reduced by 20 gigatons of CO_2 equivalent ($GtCO_2$), i.e., 68 % of the emissions in the reference scenario, in 2050, if all the actions are applied appropriately.

In practice, on the other hand, it should be bear in mind that we need the smart strategies to meet the LCS pathways in each country depending on each development stages. For that purpose, knowledge sharing becomes important. It should be noted that the actions presented in this report are not the only pathway to achieve an LCS. The important point is to use this report to encourage discussions among stakeholders and to develop specific actions for each country or region in Asia.

Keywords Asia • Greenhouse gas • Low-carbon society • Transportation sector • Building sector • Industry sector • Renewable energy • Scenario • Global computable general equilibrium model

Key Messages to Policy Makers

- GHG emissions in Asia must be reduced drastically to meet the 2 °C target, which represents that the global mean temperature should be below 2 °C compared with preindustrial level.
- This paper presents common ten actions to achieve the low-carbon society in Asian countries although their situations are quite different.
- By applying the ten actions, Asia can reduce 68 % of GHG emissions in 2050 compared with the reference scenario.
- In practice, knowledge sharing among the countries is essential to achieve leapfrog development.

1.1 Introduction

Greenhouse gas (GHG) emissions from Asia have increased continuously and accounted for approximately 38 % of global emissions in 2005 (Fig. 1.1). Considering the rapid economic growth expected in the coming decades, emissions from Asia in 2050 are projected to double the 2005 levels if efforts are not made toward achieving low-carbon societies (LCSs). The Fifth Assessment Report of IPCC Working Group III (IPCC 2014) mentioned that, in order to achieve the 2 °C target, which is to limit the increase in global mean surface temperature to less than 2° C, the GHG emissions in 2050 and 2100 will have to be reduced by 41–72 % and 78–118 %, respectively,

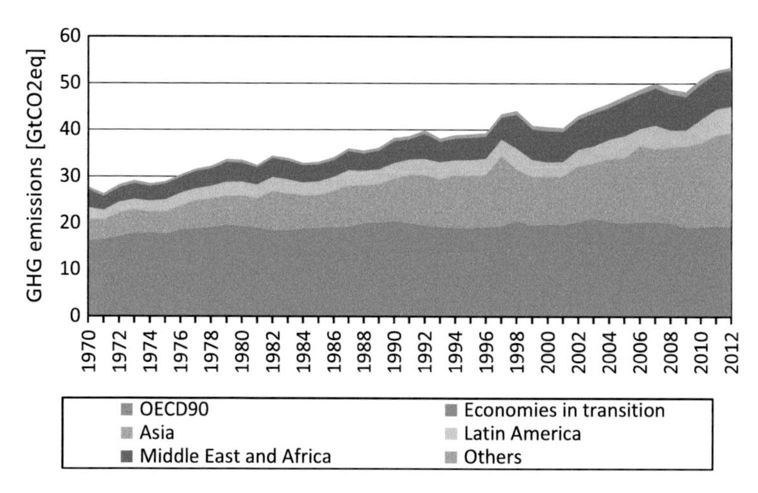

Fig. 1.1 Total GHG emissions by region between 1970 and 2012 (Notes: Data source is EDGAR v4.2 FT2012 (EDGAR 2014). Others include international aviation and shipping)

compared with the 2010 level. This means that the reduction of GHG emissions in Asia is imperative for the transition by 2050 to an LCS worldwide that has halved GHG emissions. As the energy consumption is expected to grow continuously with economic development, the reduction of CO_2 emissions from fossil fuel burning is an important goal. In addition, as the GHG emissions other than CO_2 emissions from fossil fuel burning account for approximately 40 % of the Asian GHG emissions, it is equally important to reduce them by actions like stopping deforestation, increasing CO_2 absorption from forestry, and decreasing such emissions from farmland and livestock. Furthermore, taking measures toward the realization of an LCS may also lead to the resolution of other key developmental challenges such as improving energy access, reducing local pollution, and eradicating poverty.

The LCS transition by Asian countries will not be an easy task. In order to accomplish this transition, it is vital that stakeholders including central and local governments, private sector enterprises, NGOs and NPOs, citizens, and the global community tackle it with a focused and common vision of the society they wish to achieve, while cooperating with one another and being aware of the roles they need to play.

In addition, careful attentions should be placed on the diversity of the Asian countries when it comes to the implementation of countermeasures. Depending on the country or region in Asia, the level of development, amount and type of resources, climate conditions, culture, and other factors differ, and the actions that are effective may vary accordingly (Fig. 1.2). However, guidelines showing the common requirements for realizing an LCS in Asia are extremely useful when each country considers measures and strategies that are highly feasible and effective.

There are many future scenarios, and future society will be diverse and uncertain. Based on the previous studies, we summarize the future scenarios of this study into two types: one is advanced society scenario and another is conventional society scenario (Kawase and Matsuoka 2013). Advanced society scenario will accept the new social system, institution, technologies, etc., positively and proactively. On the

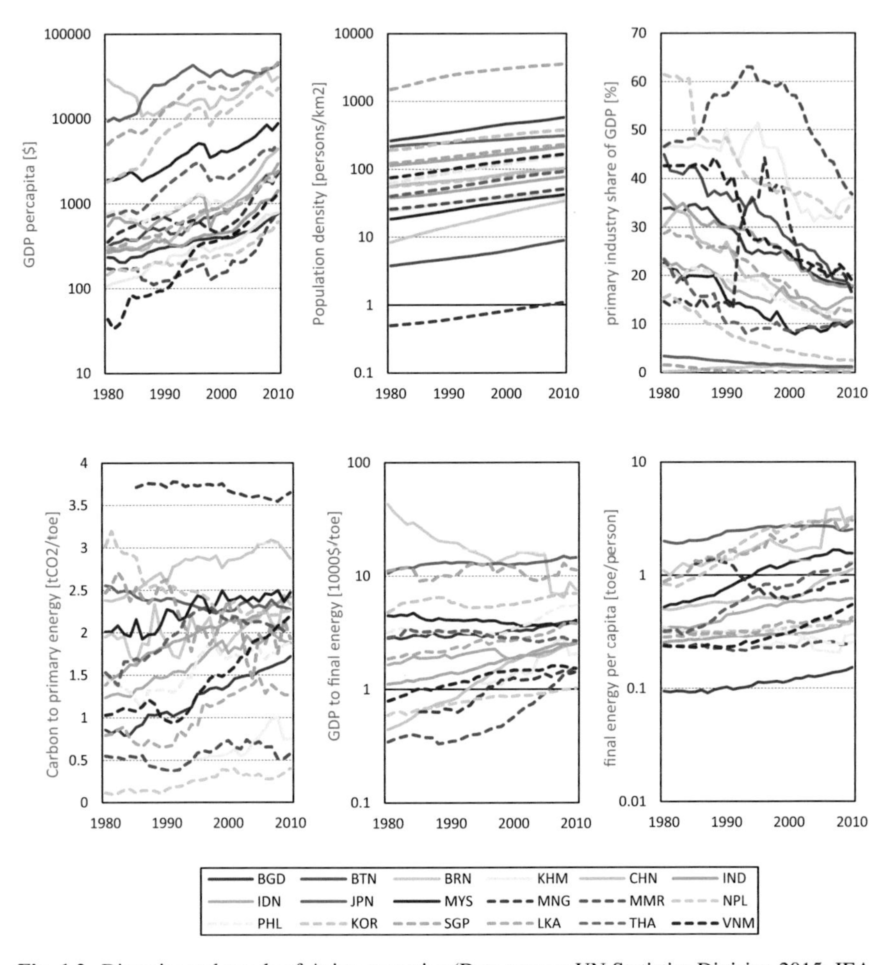

Fig. 1.2 Diversity and trends of Asian countries (Data source: UN Statistics Division 2015; IEA 2014a, b, c)

other hand, conventional society scenario will be discreet about the new social system, institution, technologies, etc., and worry about their transition cost. Table 1.1 shows the features of these two types of socioeconomic situations in Asia in 2050 for the quantitative analysis. The quantification in the following sections is based on Advanced Society Scenario.

1.2 Ten Actions to Achieve the Low-Carbon Society in Asia

In order to realize an LCS that satisfies the multifaceted needs and values of each Asian country, it is vital to gain the cooperation of a wide range of stakeholders, including policy makers, international aid agencies, private companies, local

Table 1.1 Assumptions of society in 2050

	Advanced society scenario (ADV)	Conventional society scenario (CNV)
Summary	Accepts the new social system, institution, technologies, etc., positively and proactively	Discreet about the new social system, institution, technologies, etc., and worries about their transition cost
Economy	Annual growth rate from 2005 to 2050, 3.27 %/year (global) and 4.16 %/year (Asia)	Annual growth rate from 2005 to 2050, 2.24 %/year (global) and 2.98 %/year (Asia)
Population	Total population in 2050, 9.3 billion persons in the world and 4.6 billion persons in Asia	
Education	Education system will be improved positively	Education system will be improved normally
	Education period, from 4–12 years in 2005 to 11–14 years in 2050	Education period, from 4–12 years in 2005 to 8–13 years in 2050
How to use time	Time for working and improving career will be longer	Time for staying with family or friends will be longer
Labor	Full employment in 2075	Fixed unemployment rate to 2009 level
Government	Efficiency will be improved immediately	Efficiency will be improved gradually
International cooperation	Reduction of trade barriers and FDI risks	Gradual improvement in collaborative relationships among Asian countries
Innovation	High	Medium
Transportation	Increase of demand due to high economic growth	Gradual increase of demand
Land use	More speedy and more efficient land use change	Moderate and careful land use change

communities, and NGOs, and share their long-term visions and strategies for an LCS. "Ten Actions toward Low Carbon Asia" as shown in Fig. 1.3 provides a guideline to plan and implement the strategies for an LCS in Asia (Low-Carbon Asia Research Project 2012, 2013). It takes into account the interrelationships between individual policies and the sequence in which they should be implemented. It also discusses the necessary actions to be taken by governments, private sectors, citizens, and international cooperation agencies on a priority basis.

In the following sections, each action is explained.

1.2.1 *Hierarchically Connected Compact Cities*

Economic growth has led to rapid motorization and urban sprawl in major cities in Asia, giving rise to various problems such as traffic congestion and air pollution. Nevertheless, most developing countries lack low-carbon, sustainable city planning. Many developing countries have prioritized road development in response to growing transport demand, resulting in a vicious circle in which even greater car use is induced. Since around 2000, major cities in Asia have begun to undertake urban railway development, but so far its level is not at all adequate. Developing

Fig. 1.3 Ten actions toward low carbon Asia

countries are also far behind developed countries in terms of vehicle technologies, as advanced technologies are not currently affordable.

Strategies for *low-carbon urban transport* are to AVOID unnecessary transport demand, to SHIFT transport modes to lower-carbon types, and to IMPROVE energy efficiency in transport. These can be realized with compact cities having well-connected hierarchical urban centers (AVOID strategy), a seamless and hierarchical transport system (SHIFT strategy), and low-carbon vehicles with efficient road traffic systems (IMPROVE strategy). Moreover, it is important to integrate urban transport systems with interregional transport systems in ways that reduce traffic congestion. Taking into account the CO_2 emission target of a city in a developing country, the national government is responsible for determining the appropriate types of urban structure and urban transport network consistent with the vision of interregional transport development. To support such development, international financing for green development needs to be greatly strengthened. Newly introduced international financial assistance should actively include low-carbon transport development. On the other hand, industries are responsible for developing electrification technologies for smaller vehicles to reduce congestion and CO_2 emissions. Citizens should thus be encouraged to explore a higher quality of life by using public transport and smaller vehicles, not following the conventional path of mobility growth to larger cars.

On the development pathway through 2050, according to urban agglomeration in cities along interregional rail corridors for passenger and freight transport, low-carbon urban transport systems can be developed. These transport systems will provide reliable services to support globalized economic activities by improving the efficiency of urban freight movement and increasing the speed of urban public transport. On the other hand, as resource constraints become more serious and Asian developing countries begin to become aged societies from 2030, systems adaptable to diverse transport requirements can be developed as urban infrastructure stock.

1.2.2 Mainstreaming Rail and Water in Interregional Transport

Demand for *international passenger and freight transport* has been growing in Asian developing countries compared with other regions in the world. Although international freight transport throughout Asia has low carbon emissions because it is dominated by marine transport, truck transport has been increasing for short- and medium-distance inland movement. Demand for international passenger transport in Asia, and the accompanying CO_2 emissions, has also been increasing in line with the development of the global economy and decreases in airfares due to the expansion of routes served by low-cost carriers.

Similar to the case of urban transport in Action 1, the AVOID strategy for reducing unnecessary transport demand, the SHIFT strategy for shifting to low CO_2-emitting transport modes, and the IMPROVE strategy for improving transport energy efficiency will be effective for establishing low-carbon interregional transport systems in Asia. Regarding the AVOID strategy, we propose rail-oriented development of industries on an interregional scale, in which high-speed freight railways form industrial corridors. For the SHIFT strategy, shifting away from road transport to intermodal transport based on the development of railways and waterways is necessary. In the case of the IMPROVE strategy, CO_2 emissions from vehicles, aircraft, and marine vessels can be reduced by electrification, alternative fuels, and lightweight body design. Within the continental region encompassing the area from China to the Greater Mekong Subregion (GMS), shifting from air to high-speed rail for passenger transport and from road to rail and waterways for freight transport will be highly effective. Additional reductions in CO_2 emissions can be achieved by industrial agglomeration along the high-speed freight railway corridors, which will be effective over medium and long distances in reducing the per unit time and cost. Through the implementation of these strategies, cities in coastal areas will become connected by low-carbon transport modes centered not only on maritime shipping but also on high-speed rail. A low-carbon transport system that combines high-speed rail, local rail, and technologically advanced large trailers can be introduced within the GMS region and the inland areas of China to connect with coastal areas, creating an intermodal transport system. Furthermore, by implementing an environmental impact tax, both the cost and environmental impact will be considered while siting industrial facilities and building supply chains. This will promote the formation of industry clusters along a low-carbon, interregional transport system that is centered on the mainstreaming of rail and water.

1.2.3 Smart Ways to Use Materials That Realize the Full Potential of Resources

Because of the increasing utilization of various *raw materials* such as steel and cement for the construction of social infrastructure, the penetration of durable goods, and the rising consumption of consumables in Asian nations, it is predicted that GHG emissions associated with these materials (from mining of natural resources and processing to final materials) will increase. The ratio of GHG emissions related to the production of such raw materials to gross GHG emissions is not negligible. The possibility also exists that resources used for mitigation technologies such as solar power, wind power, fuel cells, batteries, and the like might become insufficient as these technologies come to be extensively used.

The efficient utilization of these resources is therefore indispensable to achieve a meaningful reduction in GHG emissions. To attain this, it is necessary to employ innovative manufacturing that uses minimal resources, to use manufactured products as long as possible, and to reuse by-products and wastes repeatedly. Weight reduction of products, substitution of raw materials that emit excessive carbon with alternative materials, and longer life span of products should be promoted. Discarded products should be recycled using cleaner energy and better reused.

For governments, it is crucially important to design low-carbon cities and national land based on a medium- to long-term perspective to realize long-life infrastructure. Recycling and reuse systems should be established for various goods to enhance their reuse and recycling institutionally. Studies of efficient utilization of resources should also be supported. In industries, weight reduction, substitution of raw materials, and longer life span of products should be promoted to provide the same goods and services with less resource consumption and lower environmental emissions. Simultaneously, technologies related to the recycling and reuse of products and wastes should be developed and adopted.

Citizens are expected to play an important role in reducing GHG emissions related to resource use. In particular, lifestyles that are simple from a material viewpoint but create richness should be realized and practiced. For example, people could change their residence depending on each stage of life and use long-life products that allow recycling and reuse.

In addition to the above activities, international cooperation in the development and diffusion of technologies for efficient utilization of resources will reduce GHG emissions related to resource use in Asia. Furthermore, if environmental labeling systems for internationally traded products become accepted and upgraded, it will become possible for consumers to recognize and support the efforts made by producers.

1.2.4 Energy-Saving Spaces Utilizing Sunlight and Wind

As a number of Asian countries are located in tropical and subtropical regions, the demand of cooling service in the building sector has been rapidly increasing in line with their economic development and the pursue of comfort. In addition, in countries with template and subarctic zone, the demand of heating has also rapidly increasing in addition to the demand of cooling. Therefore, it is important to conduct the measures to respond to cooling and heating services in order to make low carbon in the *building sector*. In parallel, it is also necessary to address the measures to reduce the energy consumption from the appliances in the building sector as the number of appliances has been also rapidly diffused year by year in line with Asian countries' economic growth and the expansion of the economic activities.

In order to reduce the demand of cooling and heating services, it is imperative to design the buildings which can manage sunlight and humidity by making the ventilation. In line with the characteristics of each region's climate, it is also necessary to make device for insulation and make use of sunlight in order to provide sufficient cooling and heating services as well as enhance energy efficiency. Moreover, the development of *energy efficiency* building performance standard which suits to each climate zone will also contribute to the creation of high energy efficiency space.

In parallel, it is also necessary to provide financial support, such as subsidy and low interest rate loan in order to rapidly diffuse affordable high energy efficiency cooling and heating and other appliances by activating the competition to penetrate the market about the high energy efficiency appliance. The diffusion of high-efficiency appliance will assist reducing the energy demand and energy consumption in the building sector.

It is also imperative to provide social benefit in addition to the financial ones in order to promote the low carbonization in the building sector. Objective evaluation about the low-carbon activities by each business office and household, recognition of their best practices, and prize-giving to their great contribution will encourage the proactive activities toward the low carbonization.

It is essential to develop the mechanism to evaluate objectively about the effort by each stakeholder. Visualization of each stakeholder's effort by the third party's evaluation and prize-giving will be very important measures to reward their effort and encourage their continuous and proactive effort.

1.2.5 Local Production and Local Consumption of Biomass

Biomass energy can be used directly by end users or as an energy resource in production activities like power stations or other centralized energy supply facilities. It plays a vital role in low-carbon development in rural and urban areas of Asia.

Firewood and charcoal are primary energy resources used by households for cooking and hot water supply in many Asian countries. Their use causes serious health problems. Hence, improving living environments is an important associated issue for biomass use while achieving low-carbon development.

Using biomass energy as a major energy source in low-carbon Asia is ordained on establishing sustainable biomass production and utilization systems that avoid conflict with food production and forest conservation and promoting the consumption of these biomass resources locally. The installation of such energy supply systems using woody biomass, waste, and animal biomass in rural agricultural communities having plentiful biomass resources will enhance the supply of low-carbon energy, besides improving the standard of living.

For promoting the utilization of biomass in Asia, governments need to implement land use regulations and other policies that prevent conflict among "food, forest, and fuel."

Phasing out of fossil fuel subsidies is one policy which can immediately enhance competitiveness of biomass energy. In addition to supply-side policies, there are policies and measures that encourage citizens to follow sustainable land use and forest management practices that enhance biomass production and food production, minimize harvesting of forest biomass, and prompt agro-industry to make innovations of commercial biomass resources that do not compete with food production.

Since biomass production and use are dispersed, the global-scale research and development of biomass energy resources and conversion technologies, and the transfer of technology and the best practices, is very vital to develop the supply push ahead of the development of the global biomass market. In addition, the preferential support to biomass energy through carbon finance instruments, including the carbon credits, is key to promote demand-side pull from the energy market. In these contexts, the industry can play a central role in research and development, and the government's policies and programs could support the widespread adoption of such advanced biomass resources and technologies.

1.2.6 Low-Carbon Energy System Using Local Resources

Toward the realization of an LCS in Asia, the low carbonization in energy demand and supply has a vital role. Energy-saving activities and the application of *renewables* such as *solar photovoltaic (PV)* and *wind power* are keys to a reduction of GHGs. The use of renewable energies will also improve energy access, eliminate energy poverty, and establish sustainable local energy systems.

In a low-carbon Asia, it will also be essential to make fossil fuel-based energy supply systems more efficient and to facilitate coordination between fossil fuels and renewable energy, thereby improving energy security. Similarly, creation of a "smart" energy system that integrates the energy demand side will be vital. To establish these systems, governments have to develop a medium- to long-term energy policy that provides a clear direction domestically and globally on the key goals and related targets to be achieved. Achieving these goals and targets would, in the short to medium term, need institutional interventions and policy incentives that enable the introduction of renewables and energy-efficient appliances and facilities. In the long run, i.e., beyond 2030, the market pull in the wake of declining costs would deploy these technologies even without government incentives. In some countries, where the electricity access is limited by the short supply of infrastructure, the governments would have an important role to support the infrastructure supply.

The industrial sector in Asia experiences strong competition from outside the region as well as within the region. The technological innovations such as for improving grid control systems that can integrate and use diverse sources of electric power, as well as smart grids and demand responses, are important areas to enhance competitiveness of industries. Innovative industries have new market opportunities to innovate, develop, and supply solutions which can support the consumers showing preference for low-carbon or green energy sources such as solar PV systems or preferences for energy-efficient appliances or insulation technologies; the supply-side solution responds by integrating renewable energy and energy-efficient technologies to match the consumer preferences. International cooperation will also be essential. The establishment of an Asia grid network among Asian countries should be pursued using international financing mechanisms, and uniform standards should also be promoted in individual countries, creating an infrastructure for cross-border electric power interchanges. It will also be important to share best practices from the efforts in each country to encourage the use of renewable energies and to establish local weather information-gathering systems and share knowledge about the ways to use such systems.

1.2.7 Low-Emission Agricultural Technologies

The *agriculture sector* contributed 14.3 % of global anthropogenic GHG emissions in 2004, according to the Fourth Assessment Report of the IPCC (2007). To achieve the target of cutting global GHG emissions in half by 2050, mitigation options in the agriculture sector in Asia are expected to play an important role. Some mitigation measures contribute not only to GHG reductions but also to improvements in environmental conditions such as water quality and hygiene. In addition, as the cost of agricultural mitigation options is relatively low, they are attracting increasing public attention. To implement these measures, governments need to expand social infrastructure such as irrigation for water management in rice fields and to implement manure management plants for diffusion of low-emission agricultural technologies. They should also promote the dissemination of information on highly efficient fertilizer application. In particular, a gradual shift to the management of fertilization at the proper times and quantities is required in areas with excessive reliance on fertilizers. The agriculture sector should implement low-carbon water management such as midseason drainage by paddy farmers, collection of manure, and management of fertilizer and crop residues. Additionally, the methane gas emitted from manure should be actively utilized as an energy source. New technologies need to be positively adopted with the aim of achieving compatibility between productivity improvement and reduction of emissions.

If citizens select locally cultivated or raised products, local agriculture will be activated. Moreover, the selection of agricultural products produced by low-carbon farming methods will enhance their market value.

International activities are also important to promote the international joint development of low-emission agricultural technologies aimed at improving feed, livestock productivity, paddy field management, and so on. Additionally, international certification for low-carbon agricultural products should be introduced and its dissemination promoted.

CH_4 and N_2O are the main GHGs emitted by the agriculture sector. While energy-induced emissions, primarily CO_2, were a strong focus of attention in the 1990s, emissions other than CO_2 and emissions from nonenergy sectors, particularly CH_4 and N_2O, have begun to attract more attention since then have shown that the nonenergy sectors and non-CO_2 gases can potentially play an important role in future climate change mitigation, although there is greater uncertainty in estimating CO_2 emissions from land use and CH_4 and N_2O emissions than in estimating CO_2 emissions from fossil fuels.

1.2.8 Sustainable Forestry Management

Deforestation reduces forest carbon stocks, creates soil disturbances, and increases CO_2 emissions. It causes degradation of remaining forestland and lower wood productivity and inflicts severe damage on biomass growth. It is therefore important to reduce the impact of logging and improve the maintenance of *forested areas* so as to halt forest degradation, thereby reducing GHG emissions and enhancing the function of forests as a carbon sink.

Planting of trees on land that was not previously forestland is called afforestation, while planting of trees on land where a forest existed is referred to as reforestation. The Kyoto Protocol treats both afforestation and reforestation as methods of reducing emissions under the *Clean Development Mechanism*. Carbon is absorbed by trees through photosynthesis and stocked in forests and soils.

In Indonesia, peat fire and peat decomposition are major emission sources in the land use sector. Both fire management and peatland management are necessary to mitigate these emissions, in conjunction with the suppression of illegal logging, protection of ecosystems, and reduction of poverty.

To manage fires and peatland, the government is expected to play an important role by implementing land use zoning for forest protection, stopping illegal logging and unplanned land clearance, supporting the economic independence of local people by enhancing their level of education, and introducing licenses for tree planting and land clearance to encourage sustainable land use by landowners.

The private sector is expected to conduct logging and planting operations sustainably on properly licensed land, appropriately manage fires lit for land clearance, acquire forestry management skills for appropriate logging and forestation, autonomously maintain land after logging for forest regeneration, and abstain from illegal logging and consumption of illegally logged timber.

Citizens should be encouraged to understand the importance and multiple functions of forest ecosystems and to manage forests at the local level. They can contribute to reduced emissions by selecting products made of certificated wood as much as possible and actively participating in programs implemented by the government, NPOs, international society, etc. In the area of international cooperation, it is important to establish international systems to certificate sustainable management of biofuel and wood production and to regulate the importation of products that do not meet the criteria. Additionally, promotion of international cooperation for forestation and capacity development in timber-producing areas is required.

1.2.9 Technology and Finance for a Low-Carbon Society

To achieve LCSs in Asia as rapidly as possible, existing *low-carbon technologies* must be deployed and commercialized, and innovative new technologies must be developed. For these things to happen, national governments need to establish an environment for the industrial sector to invest with confidence in innovative research. They also need to create frameworks in Asia at the regional level in which each country's private sector can develop efficient technologies that will play a key role in the development of low-carbon products and deliver these products to the general public. At the present time, however, many institutional, economic, financial, and technological barriers exist that are preventing technology transfer and technology diffusion. Many studies in Asia have found that these barriers differ significantly by country and technology.

In China, India, and Thailand, for example, technologies such as wind power and bioenergy electricity production that are ready for diffusion and technology transfer for commercial use may encounter such barriers as high patent acquisition costs or a lack of local expertise with regard to imported technologies and lack of know-how and skills for their operation and maintenance. For technologies such as LED lighting or photovoltaics that are ready for diffusion and technology transfer for business or consumer use, the barriers may be the small size of the market and an exceedingly small amount of investment from overseas. Because these barriers differ depending on the stage of technological development, level of diffusion, or stage of technology transfer, governments need to consider what funding, technology policies, and support programs might be required, depending on the stage of the technology life cycle. They also need to implement this in collaboration with the private sector and the relevant international bodies.

The pool of private sector funding available holds the key to the early transfer and spread of low-carbon technologies. In the past, under the UN Framework Convention on Climate Change (UNFCCC) and the Kyoto Protocol, the Global Environment Facility acted as an interim funding institution, providing funding to developing countries. After three funds were set up (the Special Climate Change Fund, the Least Developed Countries Fund, and the Adaptation Fund), however, many issues arose including a shortage of funds and the need to determine priorities for the limited funds available. To overcome this situation, the Green Climate Fund was established in 2010 under the Copenhagen Accord of 2009, and it was stipulated that developed countries were to supply 100 billion US dollars every year until 2020. However, this amount represents a huge jump in public funding, and the search is still on to find ways to secure the funds. Exacerbating this situation is the disappointing progress of multilateral negotiations under the UNFCCC. Because urgency is required in the Asian region – as it is undergoing rapid economic development – it is necessary to consider ways to procure funds especially in this region, without waiting for further progress in multilateral negotiations. Past levels of aid from developed to developing countries cannot meet the level of funding required for the spread of technologies and products. In addition to the funding provided by developed countries under the UNFCCC and official development assistance, there is a need to find ways to mobilize diverse sources of public and private funding in the Asian region.

1.2.10 Transparent and Fair Governance That Supports Low-Carbon Asia

For Asian countries to become LCSs and enjoy the related benefits, all actors – governments, industry, citizens, and international society – need to share a common vision and strategy for an LCS. It is essential to plan, implement, and evaluate the options, with coordination of each of the respective roles.

In the past, in order to achieve the GHG emission reduction targets allocated under the Kyoto Protocol, a variety of related policy frameworks were established, and there was much discussion about the roles of national governments in implementation. To truly create LCSs, however, we cannot avoid the need for reallocation of resources and burdens in the domestic context. However, political interests can become a major factor in some cases, and it becomes difficult for national governments to plan and implement effective policies. Furthermore, due to rapid economic development, GHG emissions from developing countries – which are not under legally binding obligations to reduce emissions – are rising significantly. It will not be possible to limit the global temperature increase to 2 °C if discussions and efforts continue at the current pace for achieving emission reduction targets that were adopted based on the concept of equity when the UNFCCC entered into effect. The answer to the question of what is a fair reduction varies significantly depending on a country's perspective of what is "fair." Thus, for "*low-carbon governance*" that will achieve large, long-term reductions in GHG emissions in order to achieve the 2 °C target, national commitments are important, but it is also important that other nongovernmental stakeholders make voluntary commitments, depending on their ability to do so. Also, it will be important to create institutional designs that will allow mainstreaming of low-carbon policy, in an integrated way, of the frameworks that have so far been built on a sector-by-sector basis. And, based on them, it will be important to create efficient administrative management frameworks.

Notably, many Asian countries have formulated action plans to become LCSs, but in many cases the plans are not being implemented, or, even if they are being implemented, the effects are limited. In some cases, government fraud or corruption due to inadequate legislation or governance results in a failure to effectively utilize physical, economic, and human resources. Also, due to inadequacies in governments' management philosophy or concepts, it is not uncommon to see redundancy of policies and measures by different government ministries and agencies or inadequate sharing of information.

In this context, as a national-level initiative to establish LCSs in Asia, it is necessary to build the foundations of transparent and accountable government and to institute corruption prevention measures in the public sector, including central and local (municipal) governments. Meanwhile, the international community is expected to provide support to accelerate those efforts at the national level. For example, the World Bank and other institutions have developed frameworks for country-specific evaluations of public sector policies and institutions, and attempts are being made to reflect these efforts in their international assistance. Thus, strengthening the role of the international community in encouraging improvements in public sector management in Asian countries could be a major step forward to implement policies and measures proposed under Actions 1 through 9 of this document.

Also, as described below, Asian countries are characterized by the diversity of their political systems, and they need to plan and implement policies not only for sustainable development but also other development objectives, such as reducing health problems and poverty. In many cases, the differences between countries are mainly in scale, but they have much in common. Thus, there is a need for intergovernmental policy coordination in the planning and implementation of policies that have some compatibility between development objectives and GHG emission reductions.

Regarding the public-private sector relationship, in the past there has been excessive protection of government-related and/or certain private companies. However, it is important to establish healthy public-private partnerships by establishing objective standardization and certifications.

1.3 GHG Reduction by Introducing "Ten Actions"

The contributions of the ten actions have been quantified by a global computable general equilibrium model. The model used here divides the world into 17 regions as shown in Table 1.2 and contains the categories of governments, households, and producers. The production is classified into 32 goods. The model deals with power generation technologies in detail. This report depicts the advanced society scenario developed by the Low-Carbon Asia Research Project. About the more detailed model structure, please see Fujimori et al. (2012, 2013). Figures 1.4, 1.5, and 1.6 show the trajectories up to 2050 in population, GDP, and primary energy supply by region in the reference scenario, respectively. First, the GHG emissions in the reference case of the advanced society scenario in Table 1.1 are estimated. Then, the LCS scenarios with the ten actions are quantified, targeting a halving of global GHG emissions by 2050. Subsequently, the emission reductions and the contribution of each action in 2050 are estimated.

Table 1.2 Regional classification in this analysis

Japan (JPN)	EU25	Brazil
China (CHN)	Rest of Europe	Rest of Latin America
India (IND)	CIS	Middle East
South East Asia + Rest of East Asia (XSE)	Turkey	North Africa
Rest of Asia (XSA)	Canada	Rest of Africa
Oceania	USA	

Note: The five gray cells are regarded as Asia

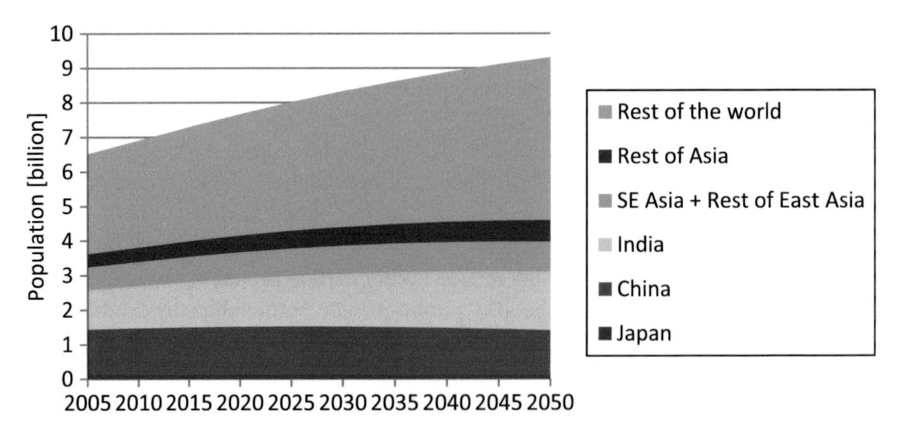

Fig. 1.4 Regional population trends by 2050 in reference scenario (unit, million)

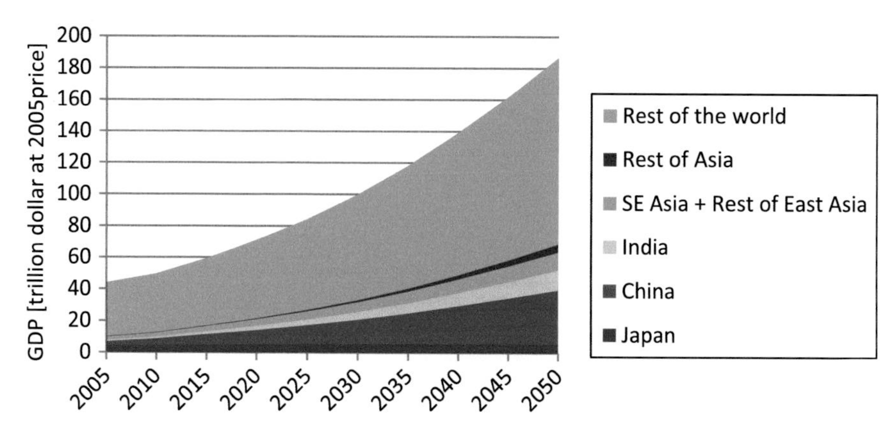

Fig. 1.5 GDP trends by 2050 in reference scenario (unit, trillion $ at 2005 price)

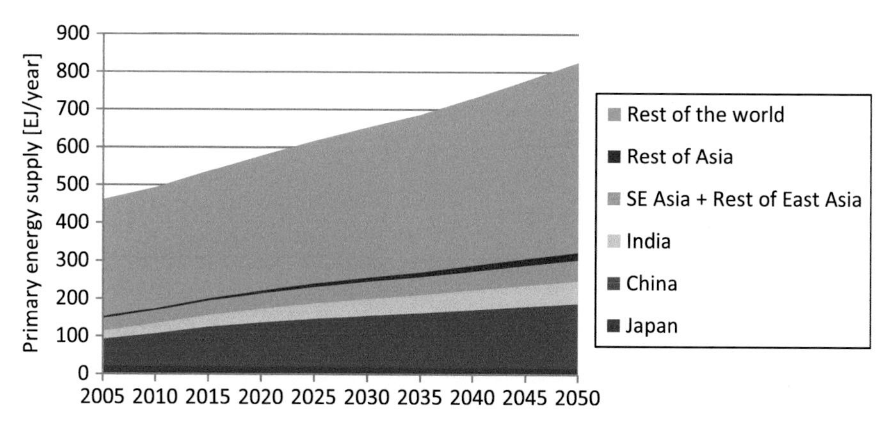

Fig. 1.6 Primary energy supply by 2050 in reference scenario (unit, EJ/year)

Figure 1.7 shows the future GHG emissions in Asia and the world in reference scenario and LCS scenario. As for the Asia, the quantities of GHG emission reduction by actions are also represented.

1.3.1 Feasibility of Reducing GHG Emissions by 68 %

If all the actions are applied appropriately, GHG emissions in Asia can be reduced by 20 gigatons of CO_2 equivalent ($GtCO_2$), i.e., 68 % of the emissions in the reference scenario, in 2050. These include all the ten actions covered in this report, and some other actions for CH_4 and N_2O emission reduction in non-agriculture sectors. Figures 1.8 and 1.9 show the primary energy supply by energy type and electricity generation by technology in Asia, respectively. From these figures, the energy saving becomes important through 2050. Moreover, introduction of

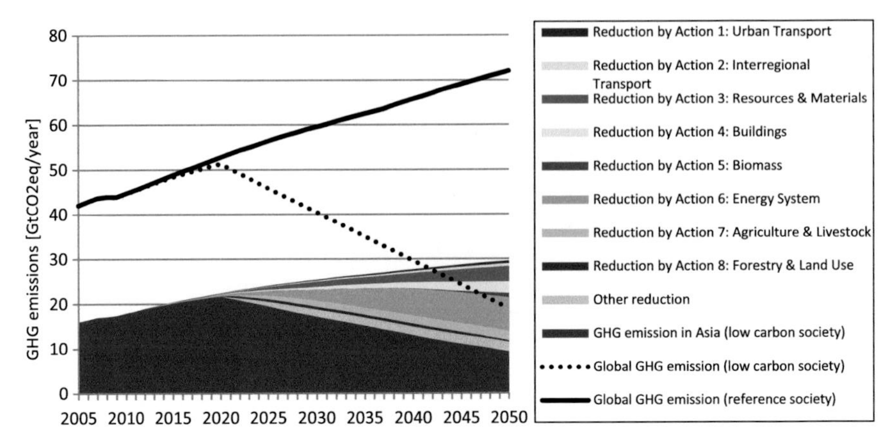

Fig. 1.7 GHG emissions in low-carbon Asia

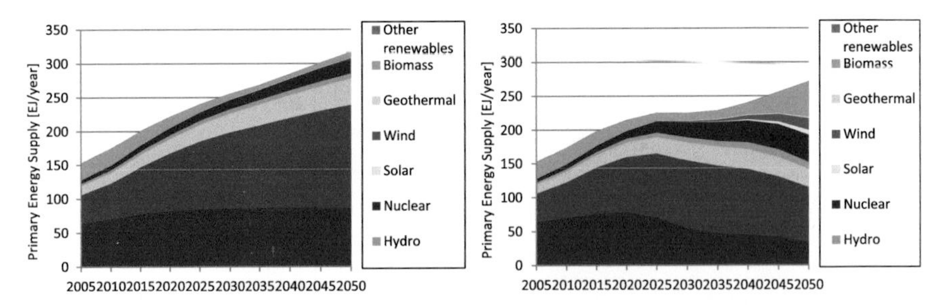

Fig. 1.8 Primary energy supply in Asia by energy type: reference scenario (*left*) and LCS scenario (*right*)

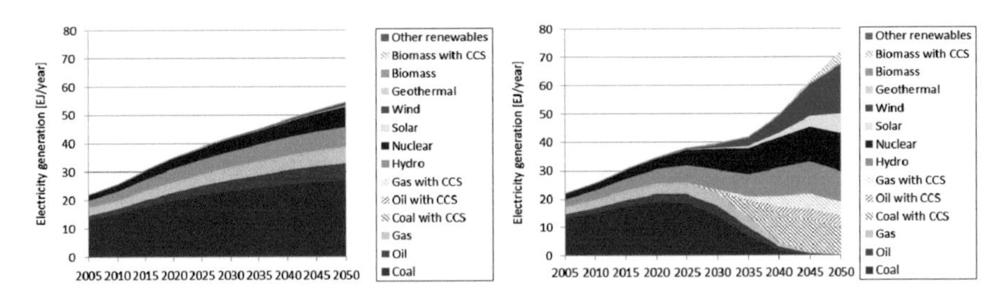

Fig. 1.9 Electricity generation in Asia by technology: reference scenario (*left*) and LCS scenario (*right*)

non-carbon energies, which include renewable energies and fossil fuels with carbon capture and storage (CCS) technology, becomes important after 2030. The share of fossil fuels in LCS scenario becomes smaller than that in reference scenario, and in the fossil fuel thermal power sectors install the CCS technology.

Actions 1 and 2, which focus on transportation, account for a combined share of 6.1 % of the total reduction in Asia. The share of Action 3, which aims to lower carbon emissions in the usage of materials, is 17 %, while the share of Action 4, which encourages energy saving in buildings, is 13 %. The share of Action 5, which utilizes biomass energy, is 4.7 %, and the share of Action 6, which is related to other energy supply systems, is 37 %. The shares of Actions 7 and 8, dealing with agriculture and forestry, are, respectively, 10 % and 1.6 %. The remaining 11 % of the reduction is accounted for by measures that are not listed in this report. The results of the actions will vary according to each country and region. For example, Actions 3, 4, and 6 will be effective for most countries and regions, whereas the contribution of Action 7 will be the largest in XSA&XOC (South Asia excluding India and small island states in Oceania) and the second largest in India.

1.4 Conclusion

As is discussed in the previous section, GHG emissions in Asia must be reduced drastically to meet the 2 °C target. In order to analyze the feasibility of such deep reduction, two scenarios are developed and analyzed in detail, namely, reference scenario and LCS scenario, and ten actions to meet the low-carbon Asia are identified. The analysis shows that it is possible to reduce the GHG emissions drastically in Asia by appropriately applying such actions. The reduction can reach 68 % from reference case. In other words, leapfrog development can be achievable in Asia.

In practice, on the other hand, it should be bear in mind that we need the smart strategies to meet the LCS pathways in each country depending on each development stage. For that purpose, knowledge sharing becomes important. It should be noted that the actions presented in this report are not the only pathway to achieve an LCS. The important point is to use this report to encourage discussions among stakeholders and to develop specific actions for each country or region in Asia.

Acknowledgment This chapter is a summary of final results of Low-Carbon Asia Research Project supported by the Environment Research and Technology Development Fund by the Ministry of the Environment, Japan (S-6-1). The authors express their thanks to all the members who contributed to this project.

References

EDGAR (2014) Global emissions EDGAR v4.2 FT2012. http://edgar.jrc.ec.europa.eu/overview.php?v=42FT2012#

Fujimori S, Masui T, Matsuoka Y (2012) AIM/CGE [basic] manual, Discussion paper series, no. 2012-01, Center for Social and Environmental Systems Research, NIES (National Institute for Environmental Studies). http://www.nies.go.jp/social/dp/pdf/2012-01.pdf

Fujimori S et al (2013) Global low carbon society scenario analysis based on two representative socioeconomic scenarios. Glob Environ Res 17(1):79–87

IEA (2014a) Energy balances of OECD countries. OECD/IEA, Paris

IEA (2014b) Energy balances of non-OECD countries. OECD/IEA, Paris

IEA (2014c) CO_2 emissions from fuel combustion. OECD/IEA, Paris

IPCC (2007) Climate change 2005: mitigation. Cambridge University Press, Cambridge and New York

IPCC (2014) Climate change 2014: mitigation. Cambridge University Press, Cambridge and New York

Kawase R, Matsuoka Y (2013) Global GHG 50% reduction and its feasibility in Asia. Glob Environ Res 17(1):11–18

Low-Carbon Asia Research Project (2012) Ten actions toward low carbon Asia. http://2050.nies.go.jp/file/ten_actions.pdf

Low-Carbon Asia Research Project (2013) Realizing low carbon Asia – contribution of ten actions. http://2050.nies.go.jp/file/ten_actions_2013.pdf

United Nations Statistics Division (2015) UNSD statistical databases. http://unstats.un.org/unsd/databases.htm

Chapter 2
Transition to a Low-Carbon Future in China Towards 2 °C Global Target

Jiang Kejun, Chenmin He, and Jia Liu

Abstract The purpose of low-carbon development in China is for both national sustainable development and global climate change action. For the global climate change target 'to hold the increase in global average temperature below 2 °C above preindustrial levels', China needs to peak in CO2 emissions by 2025 *at the latest* and then secure deep cuts in CO2 emissions. Previous studies on emission scenarios show that it is possible for China to peak in CO2 emissions by 2030 if strong policies are adopted, albeit at relatively high cost. In other words, peaking in CO2 emissions before 2025 represents a huge challenge for China. A modelling study conducted by IPAC on the 2-degree target stated that it is also still possible for China to peak in CO2 emissions before 2025 as long as several preconditions are satisfied, including optimised economic development, further energy efficiency improvements, enhanced renewable energy and nuclear development and CCS.

Energy-intensive industries consume more than 50 % of energy in China and account for more than 70 % of newly increased power output. Scenario analysis shows that many energy-intensive product outputs will reach a peak before 2020, with a much slower growth rate compared with that in the 11th Five-Year Plan, and therefore will significantly change the pathway for energy demand and CO2 emissions.

Energy efficiency should be further promoted. In the 11th Five-Year Plan, energy efficiency was improved significantly, and by reviewing what happened in this Plan compared to energy conservation efforts over the last several decades, as well as effort in other countries, it can be seen that China is now making unprecedented efforts in energy conservation. The target is to make China's energy efficiency in major sectors one of the best by 2030.

China is a now a leading country in new energy and renewable energy. Based on planning taking place in China, by 2020, renewable energy will provide 15 % of the total primary energy, which includes renewable energy excluded from the national energy statistics.

J. Kejun (✉) • C. He • J. Liu
Energy Research Institute, Beijing, China
e-mail: kjiang@eri.org.cn

S. Nishioka (ed.), *Enabling Asia to Stabilise the Climate*,
DOI 10.1007/978-981-287-826-7_2

Another key factor is the increase in natural gas use in China. In the enhanced low-carbon scenario, natural gas use will be 350 BCM by 2030 and 450 BCM by 2050; and in the 2-degree scenario, it will be around 480 BCM by 2030 and 590 BCM by 2050. Together with renewable energy, this leaves coal use in China by 2050 at below 1 billion tonnes.

For CO2 emissions, carbon capture and storage could further contribute to CO2 emission reduction. China has to use CCS if large amounts of coal are used for the next several decades, but even with the enhanced low-carbon scenario, coal use will be around 1.8 billion tonnes by 2050.

Technological progress is a key assumption for a low-carbon future for China. The cost learning curve for wind and solar and many other technologies is much stronger than the model used. Such progress greatly reduces costs in wind power and solar power within 2 years.

Keywords Emission scenario • CO2 mitigation • Modelling • Energy transition • Emission target • China

Key Message to Policymakers

- In order to achieve '2-degree' global target, China's CO2 emissions have to be at peak before 2025.
- China can peak CO2 before 2025 and reduce emission 70 % by 2050 compared with that in 2020.
- Setting a cap for CO2 emissions in China is an effective way to limit CO2 emission increases.

2.1 Background

In December, 2009, the Copenhagen Accord declared that deep cuts in global emissions are required 'so as to hold the increase in global temperature below 2 degrees Celsius'. At the climate conference in Cancun 1 year later, parties decided 'to hold the increase in global average temperature below 2 °C above preindustrial levels' and made a decision for 'strengthening the long-term global goal on the basis of best available scientific knowledge including in relation to a global average temperature rise of 1.5 °C'. The Copenhagen Accord called for an assessment that would consider strengthening the long-term goal. Further, the IPCC AR5 called on research communities to work on assessments by modelling on the emission pathway and feasibilities for the global target.

Recently, several global emission scenario studies present emission scenarios focusing on the 2-degree target, which requires global emissions to peak by 2020 at the latest (IPCC 2014). However, the commitment in the Copenhagen Accord does not agree with the global 2-degree target scenarios, which implies that further efforts are needed by each country. It is thus essential to perform more analysis at the country level to assess the potential for CO2 emission mitigation to follow the global 2-degree target pathway. This paper presents the key factors China needs to consider in order for it to follow the global target, based on modelling results from the IPAC modelling team in Energy Research Institute (ERI).

GHG emissions from energy use in China surpassed those of the United States in around 2006 and accounted for around 29 % of global emission in 2013 (Olivier et al. 2013). And due to rapid economic development, CO2 emissions are expected to increase significantly in the coming decades (IEA 2011; Kejun et al. 2009). This presents China with a huge challenge to peak in CO2 emission before 2025 and start deep cuts after 2030. Much more effort is thus required, not only in China, but by the reset of the world.

2.2 Emission Scenarios

2.2.1 Methodology Framework

In this study we used the linked Integrated Policy Assessment Model of China (IPAC) for the quantitative analysis, which covers both global emission scenario analysis and China's national emission scenario analysis. IPAC is an integrated model developed by ERI and analyses effects of global, national and regional energy and environment policies. ERI itself has conducted long-term research in developing and utilising energy models since 1992 (Kejun et al. 2009).

In order to analyse global emission scenarios and China's emission scenario, three models are used, one being global and the other two national: the IPAC-Emission global model, the IPAC-CGE model and the IPAC-AIM/technology model. The three models are linked as shown in Fig. 2.1. The modules in IPAC are currently soft linked, which means the output from one module is used as the input for another.

The IPAC-Emission model is a global model within the IPAC family and presently covers nine regions, to be extended to 22. Because this model focuses on energy and land use activities, in order to simulate other gases emissions, the model was revised to cover the analysis for HFC, PHC, SF6, CH4 and N2O.

The IPAC-Emission global model is an extended version of the AIM-Linkage model used in IPCC Special Report on Emission Scenarios (SRES) (Kejun et al. 2000), which links social and economic development, energy activities and land use activities and offers a full range of emission analyses. The IPAC-Emission global model comprises four main parts: (1) society, economy and energy activities module, which mainly analyses demand and supply under conditions of social and economic development and determines energy prices; (2) energy technology module, which analyses the short- and mid-term energy utilisation technologies under different conditions and determines the energy demand under different technology

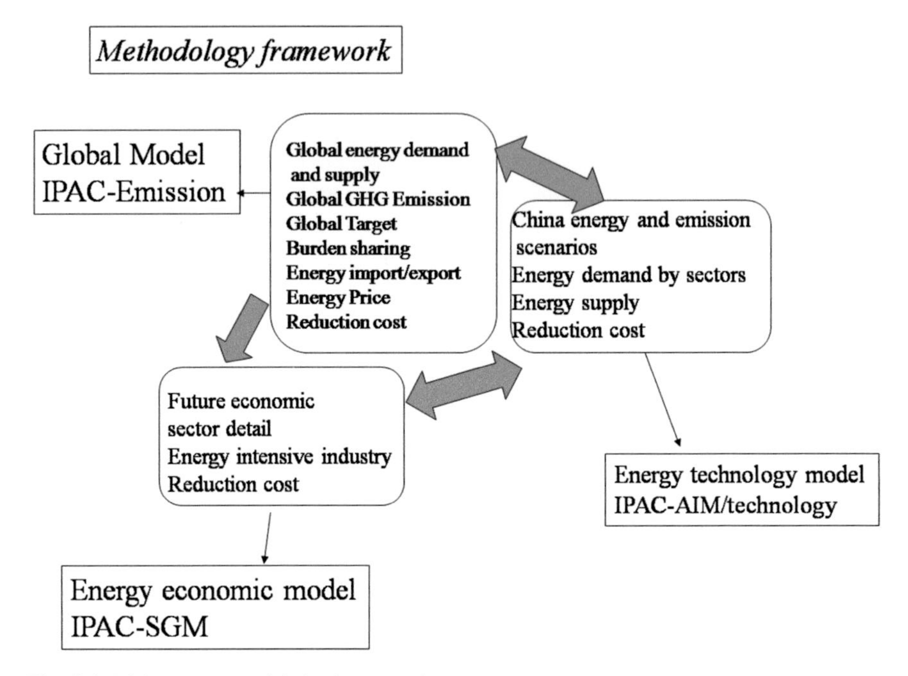

Fig. 2.1 Links among models in the research

compositions. The energy demand in this module modifies the short- and mid-term energy demand in module (1) so that the energy analysis in the macroeconomic model can better reflect short- and mid-term energy activities; (3) land use module, which analyses emissions from land use processes; and (4) industrial process emission module, which mainly analyses emissions from industrial production.

IPAC-AIM/technology is the main component of the IPAC model (Kejun et al. 1998), and it performs analyses based on the cost-minimization principle, i.e. technologies with the least costs will be selected to provide the energy service. The current version of IPAC-AIM/technology model includes 42 sectors and their products and more than 600 technologies, including existing and potential technologies.

IPAC-SGM is a computable general equilibrium model (CGE model) for China. It is mainly responsible for analysing the economic impacts of different energy and environmental policies and can analyse mid- and long-term energy and environment scenarios. IPAC-SGM divides the overall economic system into household, government, agriculture, energy and other production sectors, 42 in total.

The key focus of this study is how China can support the global 2-degree target based on its emission pathway, as well as related issues. In order to analyse the feasibility of the 2-degree target for China's emission pathway and related options, we start from the global modelling analysis performed on the target to learn how China's current emissions are affected by the 2-degree emission scenario. Then the options for the 2-degree target for China are analysed via the national model, based on the previous emission scenario analysis for China.

In the IPAC scenario setting for China, input is also needed from other relevant studies, such as GDP, population and sector outputs (Shantong 2011; Xueyi and Xiangxu 2007). The IPAC modelling team also performed studies on these parameters by using IPAC-SGM and the population model. Economic activity is becoming one of the key research topics within IPAC modelling studies due to the large uncertainty surrounding economic development and its heavy impact on energy demand. Sector development trends are crucial for energy and emission scenarios in the modelling studies, as around 50 % of total final energy use in China is accounted for by energy-intensive sectors such as ferrous and non-ferrous metal manufacture, building material manufacture and the chemical industry (China Energy Statistical Yearbook 2013 2013). In the meantime, demand for energy-intensive products was simulated by input-output analysis with a focus on downstream sector development analysis. Table 2.1 gives a scenario for energy-intensive product output in China used in the emission scenarios.

The national analysis on economic development could much more reflect national experts' viewpoints, which normally has quite big difference with the global projection on China's GDP growth. This could be seen in comparison between global modelling excises and China's national model analysis, such as IEA's World Energy Outlook (WEO) and IPAC model (IEA 2011; Kejun et al. 2009). And the sector study for output analysis could present much more sight inside economy structure change, to think about the contribution for lower energy demand and emission from economic structure change.

2.2.2 Global Emission Scenarios and Regional Allocation

The global emission scenario from IPAC mainly comes from the IPAC-Emission global model, with recent studies focusing on global mitigation scenarios. Here, a 2-degree scenario was developed based on the IPAC 450 ppm emission scenario model, as shown in Fig. 2.2. A simplified climate model, MAGICC, was used to set up the CO2 concentration at 450 ppm by year 2100 (Wigley and Raper 2001).

Regional allocation of emissions from the global emission scenario was given by using the 'burden-sharing' method. There are several ways to share the burden of emission reduction, and the subject itself has attracted much political discussion as regards whether to base it on emission per capita convergence, accumulated emission per capita convergence or something else. Sidestepping the politics, the method we used is the widely used model based on the per capita emission convergence method. With the global emission scenario from the global model, regional emission allocation was performed using CO2 emission per capita convergence criteria.

Table 2.1 Production of selected energy-intensive products

	Unit	2005	2020	2030	2040	2050
Iron and steel	10^8 tonnes	3.55	6.7	5.7	4.4	3.6
Cement	10^8 tonnes	10.6	17	16	12	9
Glass	10^8 weight cases	3.99	6.5	6.9	6.7	5.8
Copper	10^4 tonnes	260	700	700	650	460
Aluminum	10^4 tonnes	851	1600	1600	1500	1200
Lead and zinc	10^4 tonnes	510	720	700	650	550
Sodium carbonate	10^4 tonnes	1467	2300	2450	2350	2200
Caustic soda	10^4 tonnes	1264	2400	2500	2500	2400
Paper and paperboard	10^4 tonnes	6205	11,000	11,500	12,000	12,000
Chemical fertilizer	10^4 tonnes	5220	6100	6100	6100	6100
Ethylene	10^4 tonnes	756	3400	3600	3600	3300
Ammonia	10^4 tonnes	4630	5000	5000	5000	4500
Calcium carbide	10^4 tonnes	850	1000	800	700	400

Source: Author's research result

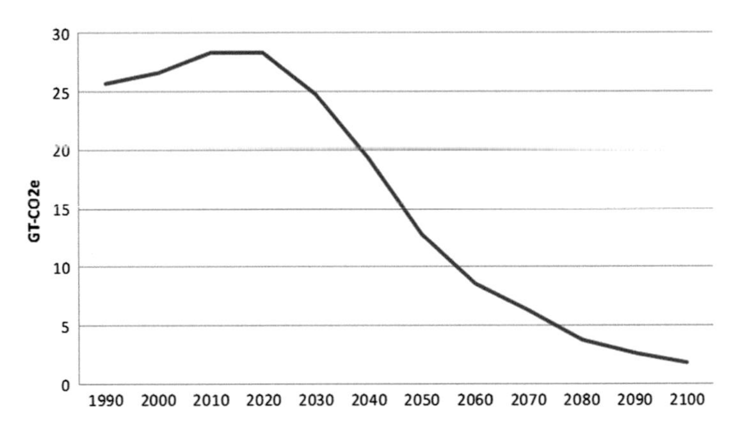

Fig. 2.2 Global CO2 emissions at 450 ppm by 2100 (Source: Author's research result)

When burden sharing using emission per capita is analysed, certain assumptions are made:

- Year to reach emission per capita convergence: here, we use 2070.
- Annex 1 countries will start reduction based on the Kyoto commitment and then proceed to deep reductions. Non-Annex 1 countries will start to depart from baseline emissions from 2010.
- CO_2 emission per capita in some developing countries may exceed developed countries.
- Population in IPAC model comes from IIASA analysis. Figure 2.3 gives the CO_2 emissions by major regions and countries.

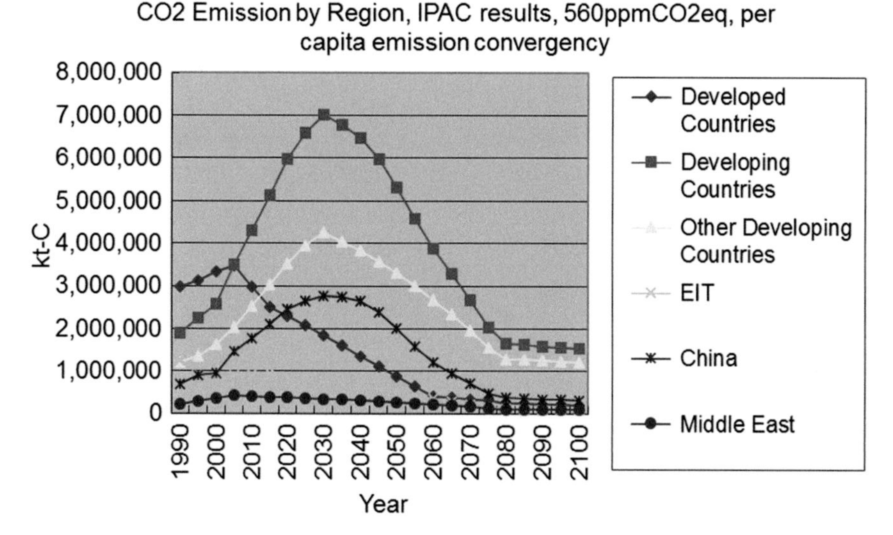

Fig. 2.3 Emissions in regions based on per capita emission convergence burden sharing (Source: Author's research result)

In order to allow more leeway for the emissions of developing countries in the future, developed countries need to make deep reductions as soon as possible. In the analysis, we also assumed other developing countries will do their part in CO2 mitigation, based on country developments and international collaboration.

The technological feasibility was also considered, which was based on the global emission scenario study from IPAC model. Figure 2.3 presents a picture for emission reduction in 2020 towards the 2-degree target.

2.2.3 China's Emission Scenarios

The IPAC team developed and published emission scenarios for China (Kejun et al. 2008, 2009), which comprise the three scenarios: baseline, low carbon and enhanced low carbon. The enhanced low-carbon (ELC) scenario involves China peaking in CO2 emissions by 2030 and then starting to decrease after that.

From Fig. 2.3, we can see China's CO2 emissions peaking at around 2025 at 8.56 billion tonnes, in order to reach the global 2-degree target. This is tougher than the enhanced low-carbon scenario from IPAC. With the assumption on GDP, the carbon intensity from 2005 to 2020 will be in the range of 49–59 % for these scenarios, which is much higher than the government target announced.

The government target announced based on a 40–45 % carbon intensity reduction between 2005 and 2020 is realised via domestic efforts. On the one hand, it is possible for China to do better if existing policies on energy efficiency, renewable energy and nuclear energy continue over the next two Five-Year Plans but with more emphasis placed on low-carbon development and low-carbon transport and

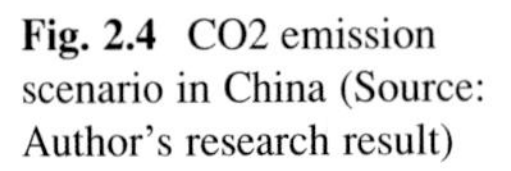

Fig. 2.4 CO2 emission scenario in China (Source: Author's research result)

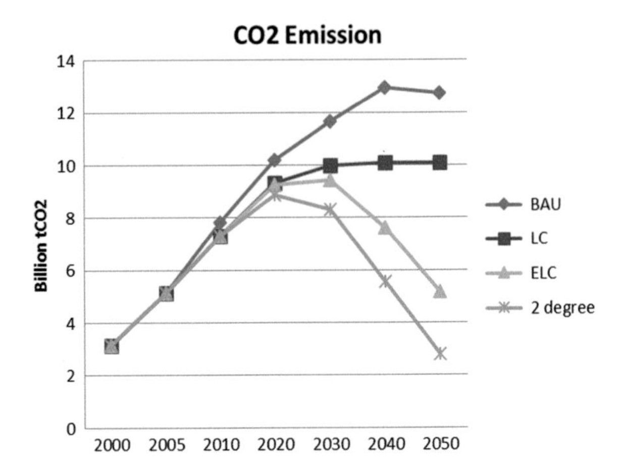

lifestyle; on the other hand, it is also possible to go further with international collaboration via technology collaboration, international carbon financing, carbon market and so on. Basically, the possibility for China to do better is high.

In order to analyse the feasibility for China, one more scenario—the 2-degree scenario for China—was given using the same model. Under this modelling analysis, we can see economic activities, energy activities, technology progress and lifestyle change in much more detail. The 2-degree scenario was developed based on the enhanced low-carbon scenario by pursuing further action in order to assess the feasibility.

Figure 2.4 presents the results for the new scenario family.

2.3 Key Factors in the Low-Emission Pathway

In the modelling analysis, key areas for CO2 emissions include economic development optimisation, energy efficiency improvements, renewable energy and nuclear development, carbon capture and storage and change of lifestyle and consumption. Efforts in the IPAC *modelling* study were based on the possibility of key assumptions by taking a broad look at driving forces, technology, the environment, social development and so on. In the enhanced low-carbon scenario, in order to reach the peak by 2030 and then start to decrease in CO2 emissions, several key challenges have to be overcome:

Change in economic structure. There was much discussion during the scenario building with the invited economics experts, as well as reviews of related studies. The GDP growth used here is the most commonly used result obtained from economic research teams, especially concerning pre-2030. Economic structural change in the three industrial sectors also presents a middle line, based on the literature reviewed. However, there is little research quantitatively detailing structural change within secondary industry. Here, based on the available research, we

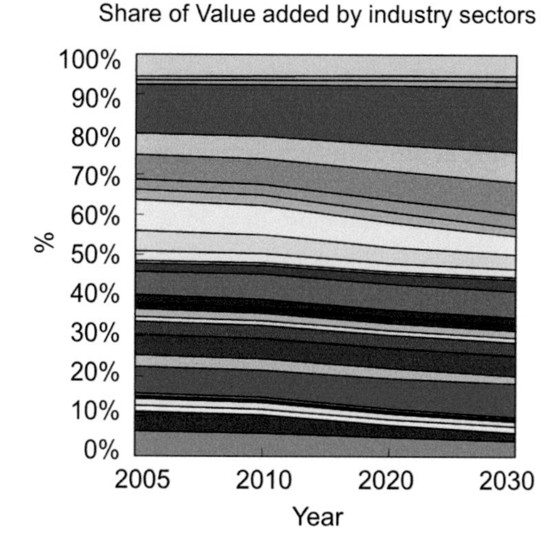

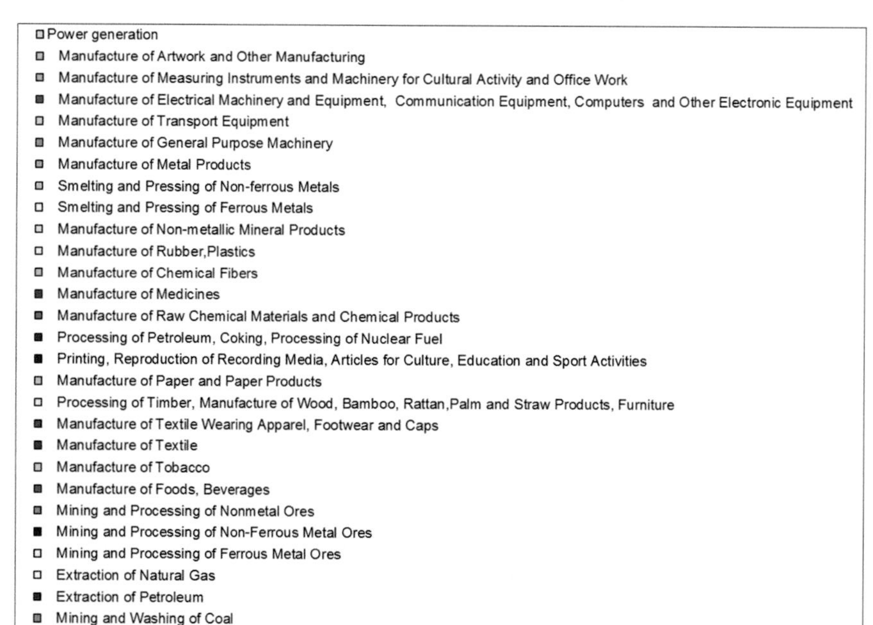

Fig. 2.5 Structural change in second industry (Source: Author's research result)

applied our own IPAC-SGM model to simulate structural change in secondary industry, as shown in Fig. 2.5.

The share of GDP from energy-intensive industry (middle part in Fig. 2.5) would reduce due to demand change. China's GDP will surpass that of the United States between 2020 and 2030, as such huge amount of GDP cannot rely on the existing economic trend involving heavy industry-driving development and raw material production. Based on a bottom-up study on the demand of energy-intensive products, it was found that many energy-intensive products will peak during 2020–2025, assuming that in the future the export of energy-intensive products will increase little, when it is already a major part of global output (see Table 2.1). This was learnt by looking at infrastructure development, including building construction, roads, railways, airports, etc., and final consumption needs, which consumed more than 95 % of cement and more than 55 % of steel (Kejun 2011). This analysis shows that the output of most of the energy-intensive products will peak before 2020.

Energy-intensive products consume nearly 50 % of energy in China, and provided that there is no significant increase in energy-intensive product production and that the growth rate thereof is much lower than GDP, the energy use increase associated with these energy-intensive products would also be limited. This would contribute greatly to decreased energy intensity per GDP and also contribute to reduced CO2 intensity.

As regards improving energy efficiency, during the 11th Five-Year Plan (2006–2010), energy efficiency was improved significantly (State Council 2011; Mark et al. 2010; Kejun 2009). In consideration of what occurred in energy efficiency during the 11th Five-Year Plan, and compared with energy conservation efforts over the last several decades and efforts made by other countries, China could be seen as having taken unprecedented action on energy conservation. Specifically, it:

- Made energy conservation policy one of the top national and top policy priorities.
- Made the energy intensity target a key indicator for local government officials.
- Involved a high number of new policies—nearly one a week from 2007 to 2008—on energy conservation from central government, in addition to local government energy policies.
- Initiated the Top 1000 Energy-Consuming Enterprise Programme, which focused on improving energy efficiency of China's largest 1000 companies which in total account for one third of China's total energy use.
- Closed small-sized power generation facilities and other industries, which was a bold measure that could have led to social unrest, unemployment and loss of profit for stockholders.

From the technical viewpoint, the above energy efficiency measures represent big achievements. China has released a total of 115 state key energy-efficient technology promotion catalogues in three batches and specially promoted seven energy-efficient technologies in the iron and steel, building material and chemical industries. Unit energy use per tonne of steel products, copper and cement decreased by 12.1 %, 35.9 % and 28.6 by 2010, respectively. By 2010, almost all advanced technologies on energy saving in industry were adopted in China. In the steel-making industry, the penetration of coke dry quenching (CDQ) increased from 30 % to more than 80 %. Use of top gas recovery turbines (TRT) increased from 49 to 597 sets. The share of furnaces with capacities above 1000 m^3 increased from

21 to 52 %. The share of new advanced rotary kilns in cement manufacturing increased from 39 to 81 %. The use of coke dry quenching in coke making increased from less than 30 % to more than 80 %. Heat recovery in cement manufacturing increased from nearly 0 to 55 %. Unit energy use for power generation supply decreased from 370 gce/kWh to 333 gce/kWh.

Owing to the widespread use of advanced high energy efficiency technologies, costs have been greatly reduced over the last several years—to the point at which some high energy efficiency technologies are even cheaper than old technologies, such as dry rotary kilns in the cement industry and super critical and ultra-super critical power generation technologies.

Such progress in energy efficiency improvements in China brings with it more opportunities for further steps in energy efficiency improvements, as follows:

- A deeper public and governmental understanding of the importance of energy efficiency. As discussed above, energy efficiency and conservation policies are one of the key issues in government—both national and local.
- Improvements in energy efficiency have been acknowledged as a means to increase economic competitiveness. Experience from other countries shows that higher energy efficiency is related to increased national economic competitiveness.
- Progress in technology towards high energy efficiency has led to new manufacturing markets for Chinese technologies. Lower cost, advanced technologies have already rapidly penetrated within China, which has profited industry. In the meantime, the international market also has a very large potential for new technologies, which will benefit not only the manufacturing industry but also energy efficiency improvements and GHG mitigation in developing countries.

It is anticipated for energy efficiency to continue improving from 2010 to 2020 in a similar manner in the 11th Five-Year Plan, based on the IPAC modelling results.

- Renewable energy development

China is the fastest-growing country for new energy and renewable energy. In order to improve the quality of the environment and promote new industry, China has extended great efforts to promote renewable energy, particularly over the past several years, and especially in wind and solar—from 2005 to 2010 the average annual growth rate exceeded 50 % annually (CEC 2011). Based on China's plans for renewable energy, by 2020 renewable energy will represent 15 % of total primary energy, which includes renewable energy not included in national statistics on energy, such as solar hot water heaters and rural household biogas digesters. Another related target is a share of non-fossil fuel energy of 15 % of the total primary energy by 2020, which includes both commercial renewable energy and nuclear energy.

- Nuclear energy development

It is expected that a nuclear energy installed capacity of over 58 GW will be realised by 2020 based on new nuclear planning, which is much larger than that original planned (40 GW).

Since the Fukushima nuclear accident in Japan, there has been much discussion on nuclear development in China; however, China has little choice in light of future power generation. Over the last several years, coal-fired power generation has increased rapidly, with an annual newly installed capacity of more than 60 GW. However, as is well known, compared to nuclear, coal-fired power generation causes high environmental and human damage. Based on the expected high demand due to energy use in China, by 2050 there is no future major role for renewable energy. Therefore, nuclear power generation will play an important role in China's future energy system by 2020.

– Carbon capture and storage (CCS)

China will have to use CCS for the next several decades if coal use continues on its present course. Even with the enhanced low-carbon scenario, coal use will be at around 1.8 billion tonnes by 2050. CCS is essential for China to enable deep cuts in CO2 emissions after 2030. Based on the study IPAC team involved for CCS implementation in China, the learning effect will have to be big to foresee the cost reduction in future. The total cost to apply CCS for 100 coal-fired power plants is not very high and will raise the price of grid electricity by 3 cent/kWh. In the enhanced low-carbon (ELC) scenario, CCS was adopted as one of the key mitigation options.

For CO2 emissions, removed CO2 emissions are given in Fig. 2.6. The key assumptions are given in Tables 2.2 and 2.3 (Kejun 2011). A lower removal rate for different power generation technologies is assumed because technological development is not yet mature at the beginning of adoption of CCS.

In the 2-degree scenario, compared with the enhanced low-carbon scenario, further implementation of renewable energy and replacing coal with natural gas were considered. For economic structural change, energy efficiency stays the same in the 2-degree scenario. Based on this, it is possible for China to peak in CO2 emissions before 2025 and then start deep cuts in CO2 emissions.

In the 2-degree scenario, renewable energy is much more extended from the enhanced low-carbon scenario. In the enhanced low-carbon scenario, power generation from renewable energy (including large hydro) will be around 34 %, and nuclear energy will account for 35 % by 2050. Installed capacity for wind, solar and hydro will be around 450 GW, 360 GW and 510 GW by 2050, respectively. In the 2-degree scenario, power generation from renewable energy could reach 48 % of the total power generation, leaving only 17 % for coal-fired power generation. Installed capacity for wind, solar and hydro is 930 GW, 1040 GW and 520 GW, respectively, by 2050.

Another key factor is the increasing use of natural gas in China. In the enhanced low-carbon scenario, natural gas use will be 350 BCM by 2030 and 450 BCM by 2050. In the 2-degree scenario, natural gas would be around 480 BCM by 2030 and 590 BCM by 2050. If natural gas is combined with renewable energy, coal use in

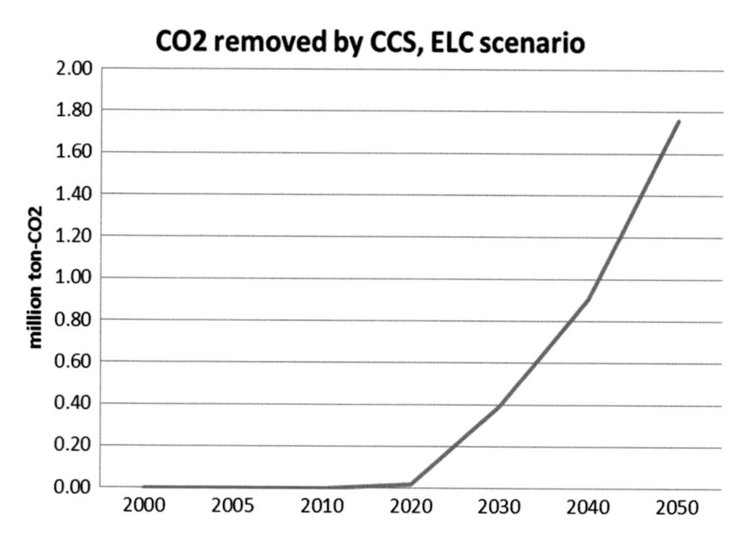

Fig. 2.6 CO2 removed by CCS in power generation sector (Source: Author's research result)

Table 2.2 Removal rate for CO2 by CCS in ELC scenario, %

	Super critical	US-critical	IGCC	IGCC fuel cell	NGCC
2020	80.0	80.0	85.0	85.0	85.0
2030	85.0	85.0	90.0	90.0	90.0
2040	85.0	85.0	90.0	90.0	90.0
2050	85.0	85.0	90.0	90.0	90.0

Source: Author's research result

Table 2.3 Power generation capacity with CCS in ELC scenario

	Super critical	US-critical	IGCC	IGCC fuel cell	NGCC
2020	0	0	1316	0	203
2030	217	379	6310	701	3411
2040	1319	2184	12,890	2275	9679
2050	2822	8465	22,045	5144	21,514

Source: Author's research result

China by 2050 will be lower than 1 billion tonnes. If so, CCS could be used for all coal-fired power plants and half of natural gas power plants.

Then, CO2 emissions in China could reach a peak before 2025, and the reduction in CO2 emissions by 2050 would be more than 70 % compared with that in 2020.

The renewable energy scenario in the 2-degree scenario is feasible owing to the recent progress in renewable energy development in China; the actual cost learning curve for wind and solar is much stronger than the model used. Technology perspective studies were also one of the key research areas in the IPAC modelling team, which has performed detailed analysis on selected technologies such as

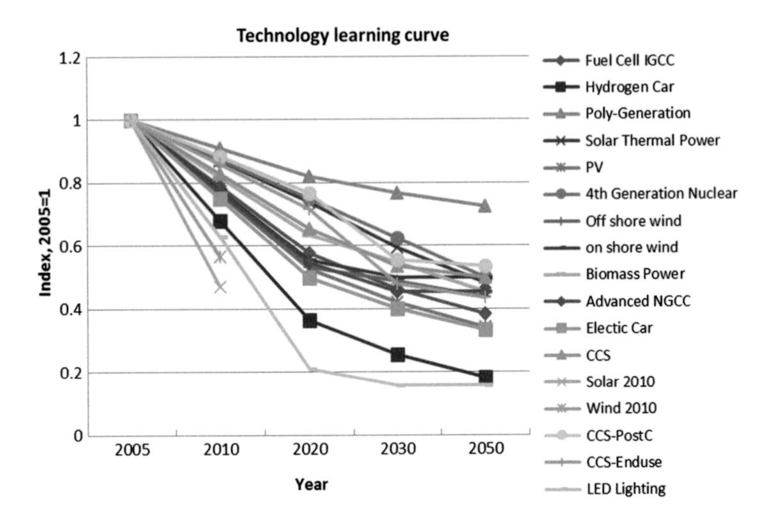

Fig. 2.7 Technology learning curve used in IPAC-AIM/technology model and data for 2010 (Source: Author's research result)

electric cars, nuclear energy, renewable energy and electric appliances (Kejun et al. 2009, 2012; Kejun 2011). Figure 2.7 presents the cost learning curve used in the model compared with actual data by 2010. Such technological progress results in a big drop in the cost of wind power and solar power within 2 years. Presently, in the coastal area, the cost of power generation for some wind farms can already compete with coal-fired power plants.

The progress in end-use technologies also moves faster than assumed by the model. Electric appliances such as LED TVs, higher-efficiency air conditioners and high-efficiency cars already had a higher penetration rate by 2011 than the model assumed. If policy is correct, a lower energy demand in the 2-degree scenario will be much more feasible by 2020 and after.

In the meantime, rapid GDP growth provides strong support for low-carbon development in China. In the 11th Five-Year Plan period (2006–2010), the annual GDP growth rate is 11.2 %, but is 16.7 % if calculated based on current value (China Statistic Yearbook 2013 2013). It is expected that by 2015, GDP in China could reach 75 trillion Yuan (at current value), newly added accumulated GDP will be 450 Trillion Yuan and cumulative GDP will be 860 Trillion Yuan. The investment needed in all modelled studies is very small compared with GDP and is normally 2–4 % or less. Regarding investment in China, new and renewable energy is one of the key sectors to be promoted within government policies and planning; thus there could be much more investment in renewable energy in the future, based on the fact that China was already the biggest investor in renewable energy as of 2010 and accounted for 24 % of the world's total.

Reviewing the progress in renewable energy planning in China, the target for renewable energy has been greatly revised upwards in recent years. Renewable

Energy Planning 2006 set the targets for wind at 30 GW and solar at 2 GW by 2020. By 2009 the National Energy Administration (NEA) announced that installed wind power generation will be 80 WG by 2020, and then in 2010 the NEA stated that installed wind power generation will reach 150 GW and solar at 20 GW by 2020. As of the end of 2011, targets for wind of 200 to 300 GW and solar of 50 to 80 GW were under discussion.

Based on the conclusion from Chinese Academy for Engineering, China's grid could adopt such renewable energy power generation in the short term.

2.3.1 Policy Options

In the modelling analysis, several policy options were simulated, one of the key ones being carbon pricing. Introducing carbon pricing, including a carbon tax or emission trading, could be an effective way to control CO2 emissions in China.

Another key policy option is setting more caps for CO2 emissions in China, which has been an effective way to limit CO2 emission increases in recent years.

2.4 Factors Causing Uncertainty in the Modelling Analysis

If we look at the scenarios, there are still several uncertainties in the emission path.

The biggest challenge is whether China's economic structure could be optimised and be directed away from a heavy industry-based and energy-intensive economy to a tertiary sector-based and less energy-intensive economy. By 2010, cement and steel output was 1.8 billion and 630 million tonnes, which is already higher than or close to the data in Table 2.2. Recently, the IPAC modelling team reanalysed the demand for cement and steel by using a methodology similar to energy forecasting, which reconfirmed the data in the table is the way for China to go. In recent years China has undergone a period of rapid infrastructure development, which cannot be sustained year-on-year going forwards. We have high confidence that many energy-intensive products will reach a peak in the near future, before 2015.

Another big uncertainty is whether the grid could adopt a large influx of renewable energy. Based on EU's experience to date, power generation from wind and solar could rise above 15 % of total power generation, and technological progress could potentially push the share of renewable energy power generation much higher (WWF 2010). However, based on the 2-degree scenario, by 2020 power generation from wind and solar in China still only accounts for 9 %.

2.5 Conclusions

If the global 2-degree target is to be implemented, China's CO2 emissions have to peak before 2025.

By using a detailed analysis modelling tool, it has been found that China could peak in CO2 emissions before 2025 and start deep cuts after that to a 70 % or greater cut by 2050 compared to 2020.

Meeting the 2-degree goal within the next 40 years will be challenging enough, and a reduction of such magnitude would require the near-simultaneous and successful deployment of all available low-carbon energy technologies and a high level of international cooperation. China will need to substantially exceed the government target announced in Copenhagen, but it is feasible if sufficient domestic action is taken and international collaboration takes place, and progress is made in technology. China's low-carbon development planning and effort should be encouraged in the future; a well-designed international regime aiming at a low-carbon pathway should be designed.

The study focus on a deep-cut emission scenario by region based on efforts and technological feasibility should be presented to show a possible future for reductions towards a 2-degree global target.

Renewable energy development policies are crucial for China to reach the 2-degree target; as with technological progress, much more renewable energy could be utilised in China. Further, China's energy system has to be diverse, and nuclear energy is still an important option due to its relative safety and low environmental impact, despite the recent developmental slowdown caused by the accident in Japan.

Carbon pricing could be introduced in the near future. It is hard to reflect shorter-term change but needs more policy support to make technology development.

Setting a cap for CO2 emissions in China has been an effective way to limit CO2 emission increases over recent years. China is now implementing cap setting on energy demand in its 12th Five-Year Plan, together with a target for non-fossil fuel energy by 2020, which will represent a good practice as regards setting up caps on CO2 emissions post-2015. In the meantime, China is implementing domestic emission trading in pilot cities and provinces that will be capped for emissions in the near future.

Specific policy recommendations are as follows:

- Place a high emphasis on optimising economic development. For a long time, China has announced its desire to adjust the economic development pattern away from heavy industry-based development to a service industry-based economy. However, little effective action has been taken. The newly announced 12th Five-Year Plan sets a GDP growth target of 7 %, which implies economic optimisation will occur. Recent government action favouring a lower economic development growth rate has started to produce results, and this action should be continued in the long term.

- Put in place a clear long-term target for CO2 emissions with specific total amount control (emission caps). China is currently attempting energy total amount control, which will provide a good basis for setting a target for total CO2 emission amount control. In this regard, setting long-term targets for CO2 emissions up to 2030 and 2050 would send a clear message that future CO2 emission reductions are being targeted.
- Introduce a carbon-pricing regime, such as carbon tax or emission trading, in the near future, to send a carbon-pricing signal. This will help push economic optimisation in the direction of a low-carbon economy.
- Make energy efficiency efforts deeper and wider ranging. Despite the huge achievements in energy efficiency in the 11th Five-Year Plan, there is still much more room for further action. Policies such as energy efficiency standards could be accelerated due to rapid progress in technologies.
- Make full support on renewable development, leave market for renewable energy development with support of feed-in tariff. Recently there has been discussion on limiting wind and solar energy, and this will obviously negatively impact on renewable energy development in China. There is plenty of space on the grid to adopt renewable energy in the future.
- Continue to support nuclear energy development and raise the security level of nuclear energy to provide cleaner energy. In China, nuclear power generation is still one of the cleanest and safest forms of energy supply compared to fossil fuel energy, which will continue to dominate China's energy system for decades. The strategy should be clear, involving more efforts to improve the technology.
- Initiate a pilot phase project as soon as possible for CCS in China. A plan should be made to have 7–10 CCS projects by 2020 to test the technology and make a decision on the best type. This will be crucial for expanding the utilisation of CCS projects post-2020.
- Do more for public awareness on low-carbon development; the public needs to be much more involved in low-carbon development as this could lead to reorientation of the manufacturing industry.

References

CEC (2011) Annual development report of China's power industry 2011. China Electricity Council (CEC), Beijing

China Energy Statistic Yearbook 2013 (2013) China energy statistic year book 2013. China Statistic Publishing House, Beijing

China Statistic Yearbook 2013 (2013) China statistic year book 2013. China Statistic Publishing House, Beijing

IEA (2011) World energy outlook 2011. IEA Publication, Paris

IPCC (2014) Climate change: mitigation. Cambridge University Press, Cambridge
Kejun J (2009) Energy efficiency improvement in China: a significant progress for the 11th Five Year Plan. Energ Effic 2(4):287–292
Kejun J (2011) Green low carbon roadmap for China's power industry. China Environment Science Publishing House, Beijing
Kejun J, Hu X, Matsuoka Y, Morita T (1998) Energy technology changes and CO_2 emission scenarios in China. Environ Econ Policy Stud 1:141–160
Kejun J, Masui T, Morita T, Matsuoka Y (2000) Long-term GHG emission scenarios of Asia-Pacific and the world. Tech Forecasting Soc Chang 61(2–3):207–229
Kejun J, Xiulian H, Xing Z, Qiang L, Songli Z (2008) China's energy demand and emission scenarios by 2050. Clim Chang Res Rev 4(5):296–302
Kejun J, Xiulian H, Qiang L, Xing Z, Hong L (2009) China's energy and emission scenario. In: China's 2050 energy and CO2 emission report. China Science Publishing House, Beijing, pp 856–934
Kejun J, Xing Z, Chenmin H (2012) China's energy and emission scenario with 2 degree target. Energy of China 2012(2)
Mark L, Lynn P, Nan, Z, David F, Nathaniel A, Hongyou L, Michael M, Nina Z, Yining Q (2010) Assessment of China's energy-saving and emission-reduction accomplishments and opportunities during the 11th Five Year Plan, Lawrence Berkeley National Laboratory, Publication Number: LBNL-3385E
Olivier JGJ, Janssens-Maenhout G, Peters JAHW, Wilson J (2013) Long-term trend in global CO2 emissions 2013 report, PBL publication number 500253004
Shantong L (2011) China's economy in 2030. Economy Science Publishing House, Beijing
State Council (2011) China's policies and actions for addressing climate change. State Council
Wigley TML, Raper S (2001) Interpretation of high projections for global-mean warming. Science 293:451–454. doi:10.1126/science.1061604
WWF (2010), 100% Renewable Energy by 2050, WWF
Xueyi W, Xiangxu Z (2007) Recent year's population forecast study review. Theory Reform No. 6, 2007

Chapter 3
India's GHG Emission Reduction and Sustainable Development

P.R. Shukla and Subash Dhar

Abstract India has made voluntary commitment for reducing the emission intensity of GDP in the year 2020 by 20–25 % below that in the year 2005. The Indian approach is based on delineating and implementing cost-effective mitigation actions which can contribute to national sustainable development goals while remaining aligned to the UNFCCC's expressed objective of keeping the average global surface temperature increase to below 2 °C over the preindustrial average. This chapter assesses three emission scenarios for India, spanning the period 2010–2050. The analysis is carried out using a bottom-up energy system model ANSWER-MARKAL, which is embedded within a soft-linked integrated model system (SLIMS).

The central themes of the three scenario storylines and assumptions are as follows: first, a business-as-usual (BAU) scenario that assumes the socioeconomic development to happen along the conventional path that includes implementation of current and announced policies and their continuation dynamically into the future; second, a conventional low carbon scenario (CLCS) which assumes imposition, over the BAU scenario, of CO_2 emission price trajectory that is equivalent to achieving the global 2 °C target; and third, a sustainable scenario that assumes a number of sustainability-oriented policies and measures which are aimed to deliver national sustainable development goals and which in turn also deliver climate mitigation, resilience, and adaptation as co-benefits. The sustainable low carbon scenario (SLCS) also delivers same cumulative emissions from India, over the period 2010–2050, as the CLCS scenario using carbon price as well as a mix of sustainability-oriented policies and measures.

The scenario analysis provides important information and insights for crafting future policies and actions that constitute an optimal roadmap of actions in India which can maximize net total benefits of carbon emissions mitigation and national sustainable development. A key contribution of the paper is the estimation of the net social value of carbon in India which is an important input for provisioning carbon finance for projects and programs as an integral part of financing NAMAs.

P.R. Shukla (✉)
Indian Institute of Management, Ahmedabad, India
e-mail: shukla@iimahd.ernet.in

S. Dhar
DTU – Dept. Management Engineering, UNEP-DTU Partnership, Copenhagen, Denmark

S. Nishioka (ed.), *Enabling Asia to Stabilise the Climate*,
DOI 10.1007/978-981-287-826-7_3

The analysis in the paper will be useful for policymakers seeking to identify the CO_2 mitigation roadmap which can constitute an optimal mix of INDCs for India.

Keywords Climate agreement • Sustainable development • Scenario modeling • Mitigation options • CO_2 Price • Social cost of carbon • $PM_{2.5}$ emission

Key Message to Policymakers

- India's CO_2 intensity declines in BAU yet inadequate for global low carbon goal.
- Carbon price affects energy supply side and leads to high share of nuclear energy and CCS.
- Sustainability policies reduce energy demand and enhance share of renewables.
- Low carbon policies aligned to sustainability goals deliver sizable co-benefits.
- Sustainability scenario delivers same carbon budget with lower social cost of carbon.

3.1 Introduction

India has endorsed the long-term target of limiting the temperature rise to under 2 °C (GoI 2008) and has also made voluntary commitment for reducing the emission intensity of GDP in the year 2020 by 20–25 % below that in the year 2005 at COP15 in Copenhagen. The "National Action Plan on Climate Change (NAPCC)" released by the Prime Minister's Office in June 2008 considers mitigation and adaptation actions implemented through eight National Missions (Table 3.1) to which the current government has added four more missions: wind, waste to energy for mitigation, and coastal and human health for adaptation.

The Indian approach to climate change is based on delineating and implementing cost-effective mitigation actions which can contribute to national sustainable development goals while remaining aligned to the UNFCCC's expressed objective of keeping the average global surface temperature increase to below 2 °C over the preindustrial average.

3.2 Model and Scenarios

3.2.1 Assessment Methodology and Model System

The integrated framework proposed in Fig. 3.1 falls under the earlier AIM family of models (Kainuma et al. 2003; Shukla et al. 2004). The bottom-up analysis is done

Table 3.1 Eight National Missions for climate change

Sr. No.	National mission	Targets
1	National solar mission	Specific targets for increasing use of solar thermal technologies in urban areas, industry, and commercial establishments
2	National mission for enhanced energy efficiency	Building on the energy conservation Act 2001
3	National mission on sustainable habitat	Extending the existing energy conservation building code, integrated land-use planning, achieving modal shifts from private to public transport, improving fuel efficiency of vehicles, alternative fuels, emphasis on urban waste management and recycling, including power production from waste
4	National water mission	20 % improvement in water use efficiency through pricing and other measures
5	National mission for sustaining the Himalayan ecosystem	Conservation of biodiversity, forest cover, and other ecological values in the Himalayan region, where glaciers are projected to recede
6	National mission for a "Green India"	Expanding forest cover from 23 to 33 %
7	National mission for sustainable agriculture	Promotion of sustainable agricultural practices
8	National mission on strategic knowledge for climate change	The plan envisions a new Climate Science Research Fund that supports activities like climate modeling and increased international collaboration; it also encourages private sector initiatives to develop adaptation and mitigation technologies

by the MARKAL model (Fishbone and Abilock 1981). MARKAL is an optimization mathematical model for analyzing the energy system and has a rich characterization of technology and fuel mix at end-use level while maintaining consistency with system constraints such as energy supply, demand, investment, and emissions (Loulou et al. 2004). The ANSWER-MARKAL model framework has been used extensively for India (Shukla et al. 2008, 2009; Dhar and Shukla 2015).

AIM/CGE and GCAM are top-down, computable general equilibrium (CGE), models used to compute the GDP loss and CO_2 price for the 2 °C stabilization scenario. AIM/CGE has been developed jointly by the National Institute for Environmental Studies (NIES), Japan, and Kyoto University, Japan (AIM Japan Team 2005). The model is used to study the relationship between the economy and environment (Masui 2005).

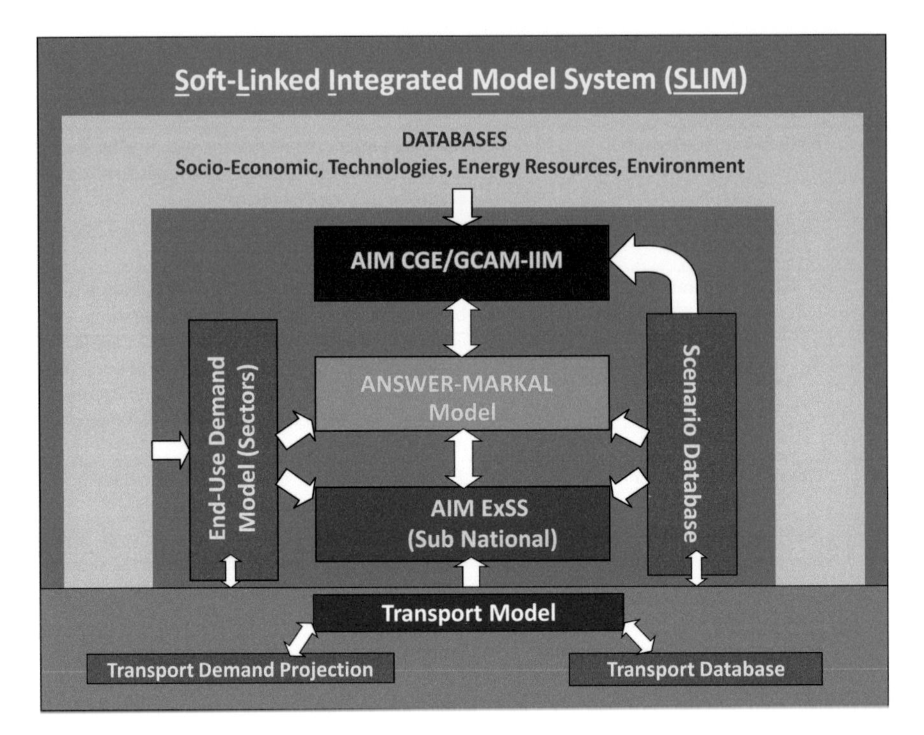

Fig. 3.1 Integrated model system

3.2.2 Scenarios Description

3.2.2.1 Business-as-Usual (BAU) Scenario

The BAU scenario considers the future economic development will copy the resource-intensive development path followed by the developed countries. The annual GDP growth rate is 8 % for the 17 years (2015–2032) and matches with the economic growth projections for India (GoI 2006, 2011). The GDP growth is expected to slow down post 2030, and the growth for overall scenario horizon, i.e., 2010–2050, is at a CAGR of 7 %. The rate of population growth and urbanization follows the UN median demographic forecast (UNPD 2013), and accordingly, the overall population is expected to increase to 1.62 billion by 2050. This scenario assumes a weak climate regime, and a stabilization target of 650 ppmv CO_2e is considered. The carbon price rises to a modest to \$20 per ton of CO_2 in 2050 (Shukla et al. 2008).

3.2.2.2 Conventional Low Carbon Scenario (CLCS)

This scenario considers a strong climate regime and a stringent carbon tax post 2020. The underlying structure of this scenario is otherwise similar to the BAU. The scenario assumes stabilization target of 450 ppmv CO_2e. The CO_2 price trajectory assumes implementation of ambitious Copenhagen pledges post 2020, and CO_2 price trajectory therefore is below 15 US $ per t CO_2 till 2020 and then increases steadily to reach 200 US $ per t CO_2 by 2050 (Lucas et al. 2013). The scenario assumes greater improvements in the energy intensity and higher share of wind and solar renewable energy compared to the BAU scenario.

3.2.2.3 Sustainable Low Carbon Scenario (SLCS)

This scenario follows the "sustainability" rationale, similar to B1 global scenario of IPCC (2000). The scenario assumes decoupling of the economic growth from resource-intensive and environmentally unsound conventional path of the BAU. The scenario seeks to achieve by significant institutional, behavioral, technological (including infrastructures), and economic measures promotion of resource conservation, energy conservation, dematerialization, and demand substitution (e.g., telecommunications to avoid travel). The scenario also considers a strong push for exploitation of large renewable energy potential (GoI 2015) and increased regional cooperation among countries in South Asia (Shukla and Dhar 2009) for energy and electricity trade and effective use of shared water and forest resources.

The scenario considers socioeconomic and climate change objectives and targets (Fig. 3.2). The SLCS considers a strong climate regime and climate objective similar to CLCS. The SLCS considers a CO_2 budget equivalent to CLCS for the period 2010–2050. However, since CO_2 mitigation is a co-benefit of a number of sustainability actions, the social cost of carbon is expected to be lower than CLCS (Shukla et al. 2008).

3.3 Scenarios Analysis and Comparative Assessment

3.3.1 Energy Demand

The overall demand for energy in the BAU is expected to increase 3.6 times from 2011 to 2611 Mtoe in 2050. The compounded annual growth rate (CAGR) is 3.6 % for the period 2011–2050 which is slower than average GDP growth of 7.0 % which has been assumed for the economy. The decoupling between GDP and energy use is due to both structural changes within the economy (greater share of service sector) and improvement in technological efficiencies. The technological efficiency

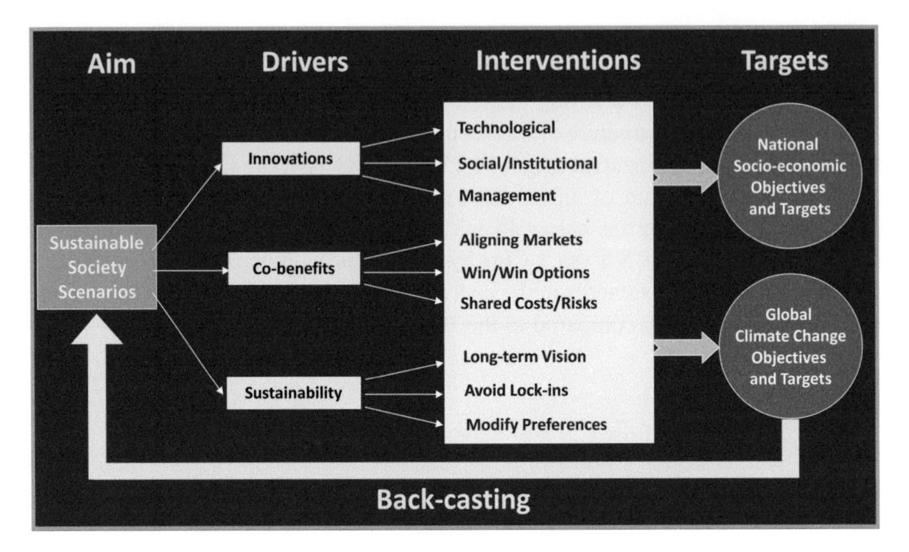

Fig. 3.2 Framework for the SLCS

improvement is most significant in the power generation where the net efficiencies improve from around 31.6 % to around 39 % in 2050.

The fuel mix is diversified in the BAU with nuclear energy, gas, and renewables taking a larger share of energy (Fig. 3.3). Coal however continues to remain the mainstay in the BAU scenario, and the bulk of coal is taken for power generation. Coal-based power generation capacity is expected to increase from 117 GW to 700 GW. Nuclear energy takes the next largest share of incremental demand for power generation, and by 2050 the installed capacity for nuclear energy is expected to increase to 200 GW from only 5 GW in 2010.

In the CLCS scenario, high carbon prices are able to bring down overall demand for energy in the medium term (by 2030); however, in the long term, the energy demand is only marginally lower than BAU (Fig. 3.4). A key reason for this is the large penetration of carbon capture and storage (CCS) in combination with coal-based power generation and steel production. CCS technology requires energy for CO_2 collection, transportation, and pumping into the storage and therefore imposes an energy penalty. The fuel mix is however diversified in a much stronger fashion with reference to the BAU, and the share of nuclear energy and renewables is much higher (Fig. 3.5).

In the SLCS energy demand is much lower (Fig. 3.4) since the demand for steel, cement, fertilizers, and many other energy-intensive commodities is much lower than BAU due to resource conservation and dematerialization. The energy demand is also lower from building, transport, and commercial sectors due to sustainable lifestyles. By 2050 the overall demand for energy is around one third lower than BAU. The fuel mix is also diversified; however, unlike CLCS, the reliance on nuclear energy and CCS is minimal and consistent with concerns with regard to their sustainability.

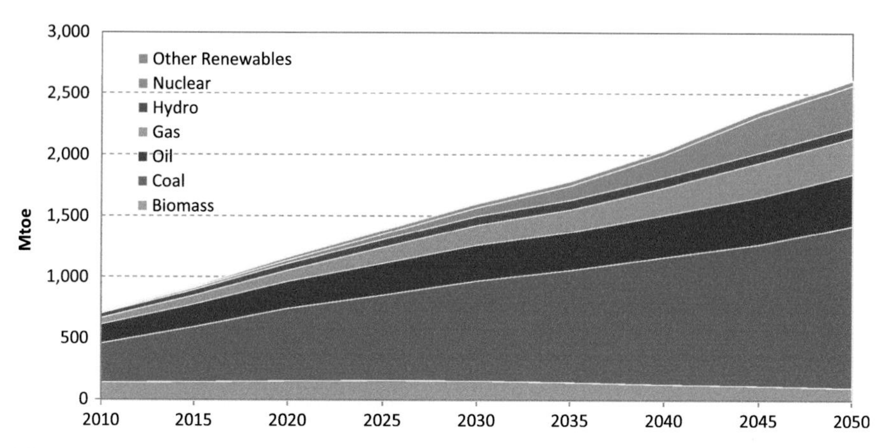

Fig. 3.3 Primary energy fuel mix and demand in the BAU

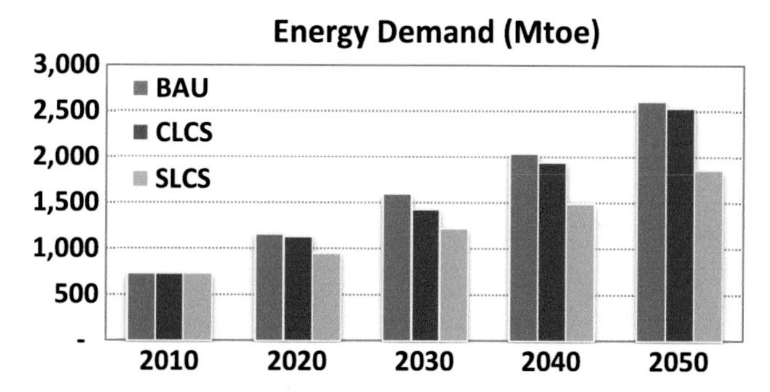

Fig. 3.4 Total primary energy demand in the BAU and low carbon scenarios

3.3.2 CO_2 Emissions and Mitigation Options

The CO_2 emissions from the energy use in the BAU increase 3.8 times between 2010 and 2050 and reach 7.32 billion tCO_2 in 2050. On a per capita basis, the emissions would be around 4.5 tCO_2 which is close to the current global average (IEA 2013). The bulk of the CO_2 emissions currently are attributable to the combustion of coal (Fig. 3.6), and this scenario would continue in the BAU in the absence of any strong climate policies.

Under both the low carbon scenarios, the growth in emissions can be limited (Fig. 3.7). In the conventional scenario, this is achieved by a small drop in energy demand (Fig. 3.4) and a sharp reduction in the share of coal from 51 % in BAU to 28 % in 2050 (Fig. 3.5). Coal is mainly substituted by nuclear energy and renewables. The share of renewable energy in 2050 is more than double from 9 % in BAU to 20 % in the CLCS (Fig. 3.5). Similarly, the share of nuclear energy is 23 % in 2050 in the CLCS. In addition coal use is increasingly decarbonized within power and steel sector with the introduction of carbon capture and storage (CCS). The total

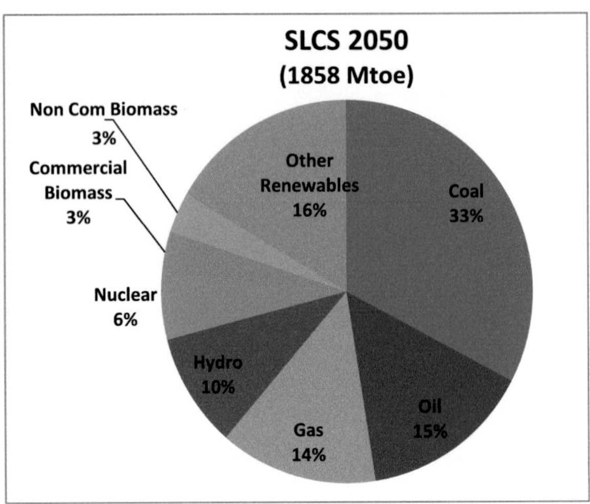

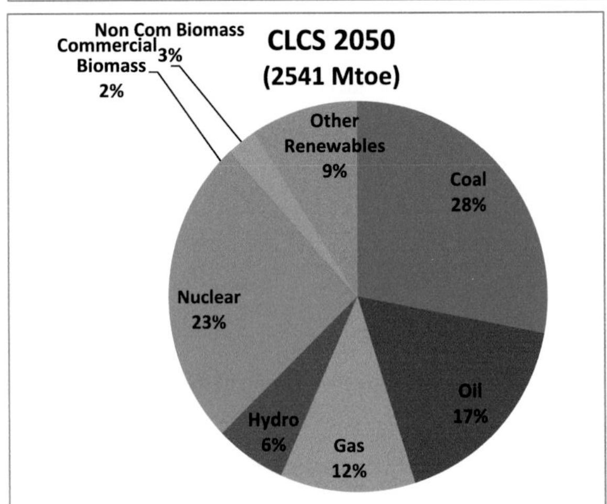

Fig. 3.5 Fuel mix in low carbon scenarios in 2050

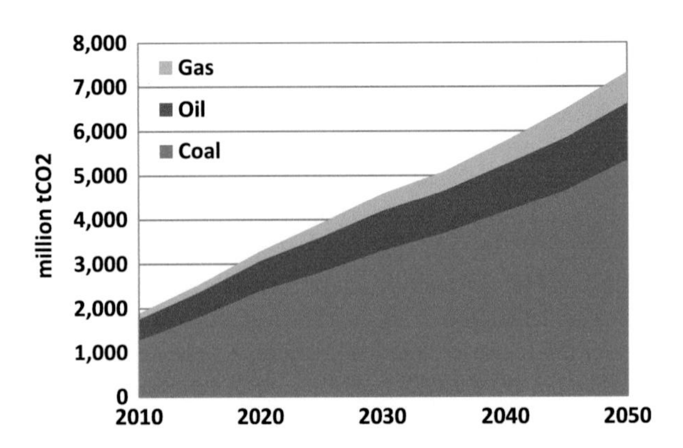

Fig. 3.6 CO_2 emissions in the BAU from energy use (million tCO_2)

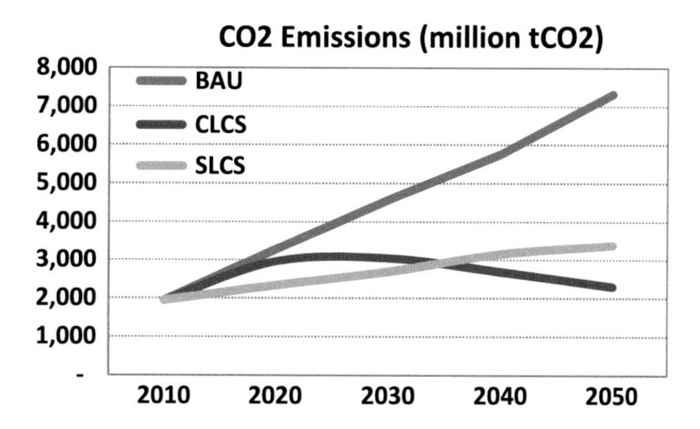

Fig. 3.7 CO_2 emissions in the BAU and low carbon scenarios from energy use (million tCO_2)

amount of CCS that is sequestered till 2050 is 30.6 billion tCO_2. A storage of less than five billion tCO_2 is available within depleted oil and gas fields and in coal mines (Holloway et al. 2009), and at many locations, this would be proximal to large point source (Garg and Shukla 2009). The supply curve for CCS therefore allows mitigation at costs below US $ 60 per tCO_2 within power and steel sector for a cumulative storage of 5 billion tCO_2. Beyond this, we have considered saline aquifers in the sedimentary basin as an option, though there is not much research or government initiative at the moment to identify potential and sites for this. Therefore, increasing CO_2 price was considered for this CO_2 storage.

In the SLCS scenario, emissions are lower due to a much lower energy demand (Fig. 3.4) from BAU. The lower energy demand is due to a wide variety of measures related to sustainability which reduce demand for energy-intensive industries like steel, cement, bricks, aluminum, etc. The second major driver is renewable energy which provides for one third of primary energy.

3.4 Co-benefits of Mitigation

Climate change mitigation can deliver co-benefits or co-costs, and we examine the scenarios on two indicators: energy security and local environment.

3.4.1 Energy Security

Energy security has been defined as the risk to the country from negative balance of energy trade and risks due to supply (Correlje and van der Linde 2006). In this sense a reduction in demand for fuel or increase in diversity of supply (Dieter 2002) is good for energy security. In terms of overall demand, the CLCS has almost similar demand as the BAU, whereas in case of SLCS, the overall demand is only 71 % of BAU in 2050 (Fig. 3.8). The fossil fuel use declines in the CLCS scenario; however,

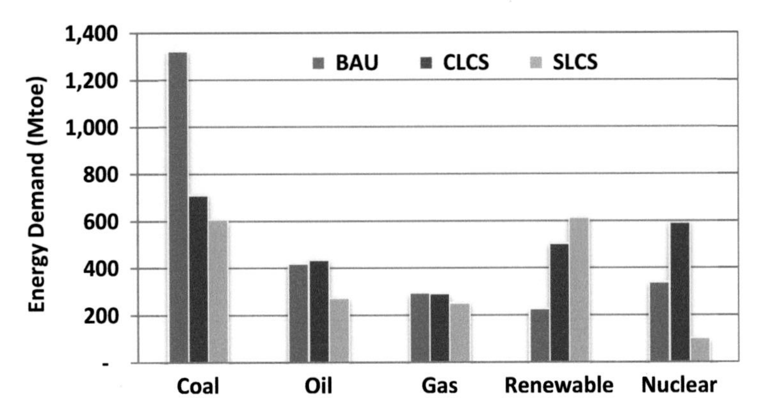

Fig. 3.8 Primary energy mix in 2050: BAU and low carbon scenarios

this is mainly due to a halving of demand for coal. Since India has a good resource availability for coal, the improvement in energy security would be small. In the SLCS scenario, the fossil fuel demand is lower for all fuels including oil, and since India depends for more than 80 % on imports of oil, improvements in energy security would be substantial. Indian nuclear energy establishment has propounded development of nuclear energy power using indigenously available thorium in the past (Kakodkar 2006); however, with signing of agreement with the nuclear energy suppliers group in 2008, India is able to import uranium. The planned nuclear energy power plants are all based on conventional fuel cycle with dependence on uranium, and therefore, higher nuclear energy will deteriorate energy security in the CLCS. In comparison the SLCS has a much lower share of nuclear energy which would help in improvement of energy security.

3.4.2 *Environment*

Many Indian cities have the very high levels of air pollution (WHO 2014) which is leading to serious health impacts (a. $PM_{2.5}$ is one of the key local pollutants and is responsible for severe health risks. Transport sector accounts for 30–50 % of the $PM_{2.5}$ (Guttikunda and Mohan 2014), and therefore, we analyze $PM_{2.5}$ for transport sector.

In India Bharat Stage III emission standard for motor vehicle (equivalent to Euro III) is applicable across India, and BS IV emission standards are applicable in the National Capital Region of Delhi and 20 other larger cities. Thirty additional cities are planned to move to Euro IV by 2015 (GoI 2014). In all the three scenarios, it is assumed that the BS IV would be fully implemented by 2020 all across India (GoI 2014).

The implementation of stricter emission norms which will entail changes to both vehicles and fuels will deliver for environment in the medium term (post 2025

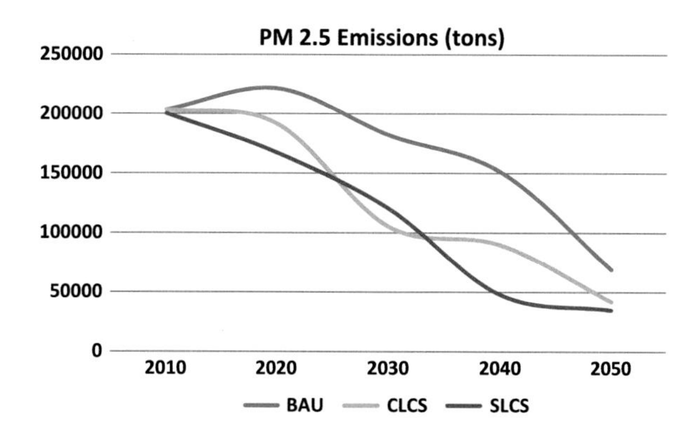

Fig. 3.9 PM 2.5 emissions from transport sector across scenarios

onwards); however, air pollution would remain a challenge for the next 10 years. However, strong sustainability measures as envisaged in SLCS can help in turning the tide on air pollution quite early (Fig. 3.9). Similarly, a strong climate regime can also bring significant benefits for air quality (Fig. 3.9).

3.4.3 *Net Social Cost of Carbon*

The CO_2 mitigation is the same between the two low carbon scenarios. In conventional scenario, the mitigation actions are mainly a consequence of a high carbon price which increases rapidly post 2020 and with an expectation of a good climate treaty in 2015. The advance measures taken as a part of the sustainability paradigm can help to put the country on a trajectory where CO_2 mitigation is a co-benefit and, because of this, the society can achieve a similar amount of mitigation at a lower social cost of carbon (Fig. 3.10). This means if sustainability is limited to India, a higher mitigation corresponding to the global carbon price will occur, which can then be traded. If the sustainability paradigm is global, then a mild tax trajectory (Fig. 3.10) is required.

3.5 Conclusions

The chapter presented historical projections of energy and emissions in India under different scenarios. The approach followed in this paper visualizes low carbon transition in India from two different perspectives. First is the conventional perspective which assumes the rest of the economy is in competitive equilibrium. The approach visualizes carbon mitigation as an outcome of the application of a globally efficient carbon price in the form of a tax or a shadow price resulting from the global emissions carbon cap. This perspective, referred to as conventional low carbon scenario (CLCS), however discounts the fact that developing country

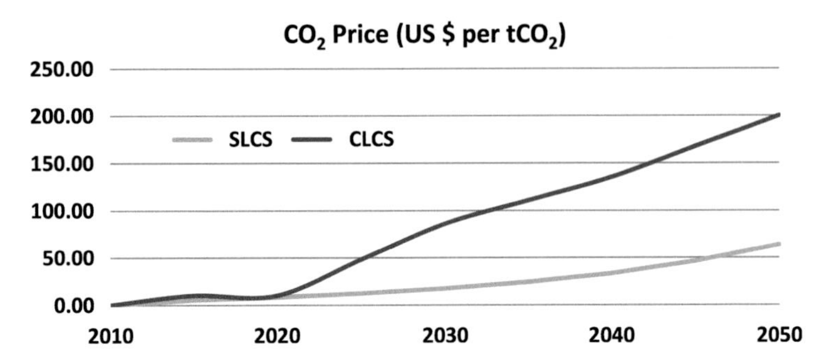

Fig. 3.10 Net social cost of carbon

economies have deep-rooted institutional weaknesses which impedes competitive behavior. The paper proposes a second scenario, referred to as sustainable low carbon scenario (SLCS), that explicitly recognizes the market weakness and hence explicitly implement additional policies which align the national sustainable development goals with the global low carbon objective.

As a reference point for the low carbon pathway, a business-as-usual (BAU) scenario is also assessed. A notable result is that energy demand and CO_2 emissions in India decouple significantly from GDP growth even in the BAU. However, the decoupling of CO_2 is not adequate when compared to what would a cost-effective global carbon regime targeting 2 °C temperature stabilization. Thus, further carbon mitigation is needed to align India's mitigation target with global stabilization.

Under CLCS, the application of global carbon price has little impact on energy demand, but it results in greater energy supply-side response like higher share of nuclear energy power and CCS. The projections show that by 2050, India can deploy nearly 30 billion tCO_2 sequestration capacity under CCS. This is much higher than what is available in depleted oil and gas wells and coal mines, and using this capacity at higher end can be extremely risky due to the uncertainty of the CCS capacity and costs in India. This aside, in this scenario, nuclear energy would supply nearly a quarter of the primary energy demand in 2050. This is also a high risk proposition given the uncertainty of the full cost of nuclear energy in India.

Under the SLCS, many sustainable development-focused measures such as designing and implementing sustainable habitat and mobility solutions, 3R (reduce, reuse, recycle) measures, and demand-side energy and resources management measures result in reducing the energy demand by a third in 2050. In addition, the policy support for renewable energy results in relatively minimal use of CCS which can be easily sequestered within the depleted oil and gas wells or coal mines in the country. The demand for nuclear energy power is also reduced significantly under this scenario. Solar and wind energy would play a bigger role in both CLCS and SLCS (Fig. 3.8). The energy security benefits, compared to BAU, are very high in SLCS but negligible in CLCS. Air quality benefits are high in both CLCS and SLCS.

In case of CLCS, the mitigation is achieved by applying the global carbon price over Indian economy. In case of SLCS, the emissions budget is assumed to be the same as the emissions in CLCS during the period 2010–2050. In SLCS, the emissions are at first reduced by various measures targeted to achieve national sustainable development goals. The budgeted carbon pathway is achieved by the shadow price of carbon corresponding to the budget constraint. This cost, which we refer to as the "social cost of carbon," is much lower in the case of SLCS since the carbon reduction that is delivered by the sustainability measures is assumed to be "free" since their cost is included in the cost-benefit assessment of national sustainability measures which typically do not include carbon benefits.

The assessment in the paper shows that aligning actions toward India's low carbon pathway with measures for achieving national sustainable development goals would result in significantly lower social cost of carbon for India. This signifies the existence of sizable co-benefits between low carbon and sustainable development actions. The methodology and analysis in this paper thus provides a way forward for scientifically delineating the Intended Nationally Determined Contributions (INDCs) for mitigation. The technological and financial details underlying the modeling analysis can be useful for preparing the road map of India's Nationally Appropriate Mitigation Actions (NAMAs) and downscale these to actionable projects with clearly identified pathways for technology development, transfer and deployment, as well as access to carbon finance.

References

AIM Japan Team (2005) AIM/CGE [Country]: data and program manual. National Institute for Environmental Studies, Tsukuba

Correljé A, van der Linde C (2006) Energy supply security and geopolitics: a European perspective. Energy Policy 34:532–543

Dhar S, Shukla PR (2015) Low carbon scenarios for transport in India: co-benefits analysis. Energy Policy 81:186–198

Dieter H (2002) Energy policy: security of supply, sustainability and competition. Energy Policy 30:173–184

Fishbone LG, Abilock H (1981) MARKAL, a linear programming model for energy system analysis: technical description of the BNL version. Int J Energy Res 5:353–375

Garg A, Shukla PR (2009) Coal and energy security for India: role of carbon dioxide (CO2) capture and storage (CCS). Energy 34:1032–1041

GoI (2006) Integrated energy policy: report of the expert committee. Planning Commission, Government of India (GoI), New Delhi

GoI (2008) National action plan on climate change. Prime Minister's Council on Climate Change (NAACP), New Delhi. http://www.moef.nic.in/modules/about-the-ministry/CCD/NAP_E.pdf. Visited on 23 Sept, 2014

GoI (2011) Low carbon strategies for inclusive growth. Planning Commission, Government of India (GoI), New Delhi

GoI (2014) Auto fuel vision and policy 2025: report of the expert committee. Planning Commission, Government of India (GoI), New Delhi. Available at http://petroleum.nic.in/autopol.pdf. Accessed 11 July 2014

GoI (2015) Report on India's renewable electricity roadmap 2030: toward accelerated renewable electricity deployment. Niti Aayog, Government of India (GoI), New Delhi

Guttikunda SK, Mohan D (2014) Re-fueling road transport for better air quality in India. Energy Policy 68:556–561

Holloway S, Garg A, Kapshe M, Deshpande A, Pracha AS, Khan SR, Mahmood MA, Singh TN, Kirk KL, Gale J (2009) An assessment of the CO2 storage potential of the Indian subcontinent. Energy Procedia 1:2607–2613

IEA (2013) World energy outlook 2013. OECD/IEA, Paris

IPCC (2000) Emission scenarios. Cambridge Universities Press, Cambridge

Kainuma M, Matsuoka Y, Morita T (2003) AIM modeling: overview and major findings. In: Kainuma M, Matsuoka Y, Morita T (eds) Climate policy assessment: Asia Pacific integrated modeling. Springer, Tokyo

Kakodkar A (2006) Role of nuclear in India's power-mix. Energy conclave 2006: expanding options for power sector. IRADe, Infraline database http://www.infraline.com/power/default.asp?idCategory=2275&URL1=/power/Presentations/Others/EnergyConclave06/EnergyConclaveConferencePresent2006-Index.asp. Downloaded on 26 Sep 2007

Loulou R, Goldstein G, Noble K (2004) Documentation for the MARKAL family of models, October 2004. 13 Sept 2007. http://www.etsap.org/documentation.asp

Lucas PL, Shukla PR, Chen W, van Ruijven BJ, Dhar S, den Elzen MGJ, van Vuuren DP (2013) Implications of the international reduction pledges on long-term energy system changes and costs in China and India. Energy Policy 63:1032–1041

Masui T (2005) Concept of CGE model and simple GE model based on IO data. In: AIM training workshop 2005, National Institute of Environmental Studies, Tsukuba, Japan

Shukla PR, Dhar S (2009) Regional cooperation towards trans -country natural gas market: an economic assessment for India. Int J Energy Sect Manage 3:251–274

Shukla PR, Rana A, Garg A, Kapshe M, Nair R (2004) Climate policy assessment for India: applications of Asia Pacific Integrated Model (AIM). Universities Press, New Delhi

Shukla PR, Dhar S, Mahapatra D (2008) Low carbon society scenarios for India. Clim Pol 8:S156–S176

Shukla PR, Dhar S, Victor DG, Jackson M (2009) Assessment of demand for natural gas from the electricity sector in India. Energy Policy 37:3520–3535

UNPD (2013) The world population prospects: the 2012 revision. United Nations Population Division, 23 Dec 2013. http://esa.un.org/wpp/unpp/panel_population.htm

WHO (2014) Ambient (outdoor) air pollution database, by country and city. World Health Organization, Geneva, Switzerland. http://www.who.int/phe/health_topics/outdoorair/databases/cities/en/. Downloaded on 01 Oct 2014

Chapter 4
Eighty Percent Reduction Scenario in Japan

Toshihiko Masui, Ken Oshiro, and Mikiko Kainuma

Abstract Toward the achievement of the 2 °C target, Japan has set several GHG mitigation targets after ratifying the Kyoto Protocol. In 2008, in order to discuss the GHG mitigation target in 2020 at COP15 held in Copenhagen, the committee on the mid- and long-term target in Japan was organized at the Cabinet Secretariat. At that discussion, the proposed six options were quantified, and finally, 15 % reduction in 2020 compared with 2005 level was selected as a target.

Because of the change of government, the new mitigation target in 2020, 25 % reduction compared to 1990, was announced at the United Nations Climate Change Summit in 2009. Then, the road maps to achieve this 25 % reduction target were quantified at the Central Environment Council. But just after they completed the road maps to achieve the target, the Great East Japan Earthquake and Fukushima Daiichi Nuclear Power Plant accident happened on March 11, 2011. Due to the nuclear power accident, the GHG mitigation road map and the future energy mix should be reconsidered. In 2012, the Energy and Environment Council forecasted that the GHG emissions in 2020 would be 5–9 % reduction compared to the 1990 level in the case of the prudent economic growth case.

After the change of government again, at the COP19 of UNFCCC in Warszawa in 2013, the new GHG mitigation target in 2020 was announced to be 3.8 % reduction compared to the 2005 level under the assumption of the no nuclear power supply. And in 2015, the mitigation target in 2030 was proposed to 1.042 GtCO2, that is, 26.0 % reduction compared to the 2013 level.

On the other hand, in the 4th Environmental Basic Plan endorsed by the Cabinet in 2012, the 80 % reduction of the GHG emissions was written clearly. In this paper, in order to assess the feasibility of this 2050 target, we utilized the AIM/Enduse to disaggregate Japan into 10 regions. The treated technologies include renewable

T. Masui (✉)
National Institute for Environmental Studies, Ibaraki, Japan
e-mail: masui@nies.go.jp

K. Oshiro
Mizuho Information and Research Institute, Tokyo, Japan

M. Kainuma
National Institute for Environmental Studies, Ibaraki, Japan

Institute for Global Environmental Strategies, Kanagawa, Japan

S. Nishioka (ed.), *Enabling Asia to Stabilise the Climate*,
DOI 10.1007/978-981-287-826-7_4

energy technologies, carbon capture and storage (CCS), and energy-saving technologies. The study shows that it is feasible to achieve 80 % emission reduction in Japan even without nuclear power. The impact of nuclear phaseout as compared to the illustrative scenario is relatively small in the long term because of the small share of nuclear energy in 2050 in any case. Achieving long-term emission reduction target proves to be still feasible with substantial increase of renewable energy, particularly solar PV and wind power. The share of renewable energy in electricity supply reaches approximately 85 % in 2050, and variable renewable energies account for about 63 % in electricity generation in 2050, hence imposing a further challenge for integration into the electricity system.

The feature of mitigation target in Japan is mainly based on the bottom-up approach. That is to say, the process stressed the feasibility of the target. On the other hand, the top-down decision is also requested for the ambitious reduction target. Toward the achievement of 2 °C target, taking actions with the long-term perspective becomes more important.

Keywords Mitigation • Carbon dioxide • 2 °C target • Energy mix • Electricity • Enduse model • Japan • Nuclear power • Carbon capture and storage

Key Message to Policy Makers

- Japan's mitigation target in 2050 is set to be by 80 % reduction by the 4th Environmental Basic Plan.
- The existing target is set based on the available technologies, that is to say, the bottom-up approach.
- Top-down decision is also requested for the ambitious reduction target.
- The options to achieve the ambitious target have been available.

4.1 Introduction

In 2013 and 2014, the IPCC 5th Assessment Reports were approved. In this report, it is concluded that "It is extremely likely that human influence has been the dominant cause of observed warming since the mid-20th century" (IPCC 2013). The IPCC 5th Assessment Report mentions that, in order to achieve the 2 °C target, which is to limit the increase in global mean surface temperature to less than 2 °C, the GHG emissions in 2050 and 2100 will have to be reduced by 41–72 % and 78–118 %, respectively, compared with the 2010 level (IPCC 2014). Figure 4.1 shows the future GHG emissions, and the right blue range (430–480 ppmCO_2eq) is the emission pathways to achieve the 2 °C target likely.

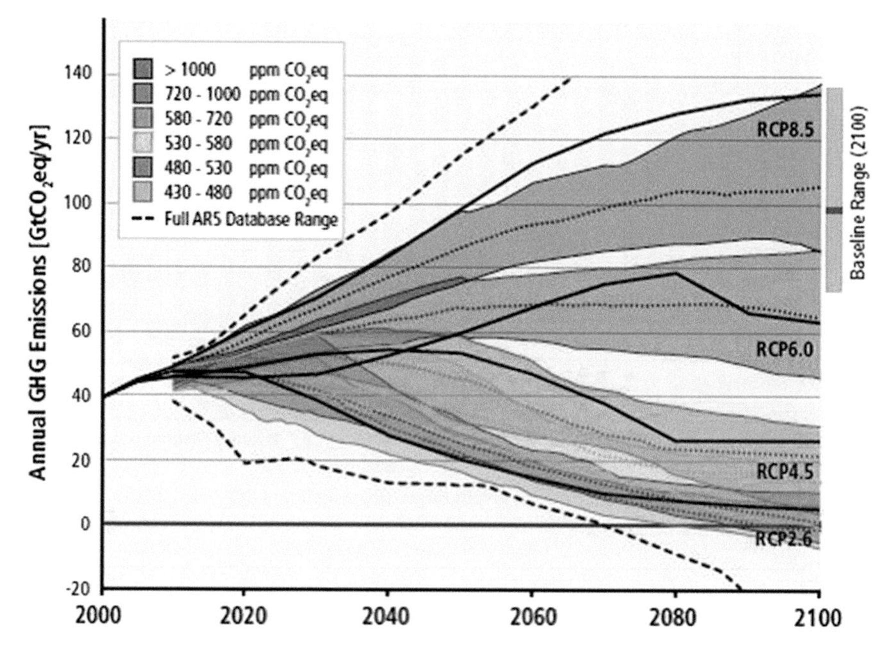

Fig. 4.1 Total GHG emissions in all IPCC AR5 scenarios

Toward the achievement of the 2 °C target, Japan has set several GHG mitigation targets after ratifying the Kyoto Protocol. In the following sections, mitigation target will be discussed and the quantification figures using the AIM (Asia-Pacific Integrated Model) are explained.

4.2 From the Kyoto Protocol to Middle-Term Target

Figure 4.2 shows the trend of GHG emissions in Japan from 1990 to 2013. The actual GHG emissions during the 1st commitment period in Japan exceeded the emission target set by the Kyoto Protocol, but if the effects of carbon sink and Kyoto Mechanism such as mitigation outside of Japan are taken into account, the emissions in Japan could achieve the emission target. On the other hand, the trend of GHG emission after 2010 would increase because of the Great East Japan Earthquake and Fukushima Daiichi Nuclear Power Plant accident in March 2011. Figure 4.3 shows the sectoral GHG emissions in Japan. Among the sectors, the emissions from commercial and residential sectors increased after the nuclear power plant accident.

In 2015, the COP21 of UNFCCC will be held in Paris. At this meeting, the post-2020 target will be discussed. Prior to the COP21, INDCs (Intended Nationally Determined Contributions), that is to say, the new ambitious target of GHG emission reduction, will be announced by each country by March 2015. Japan has been discussing this issue under the joint committee between the Ministry of the

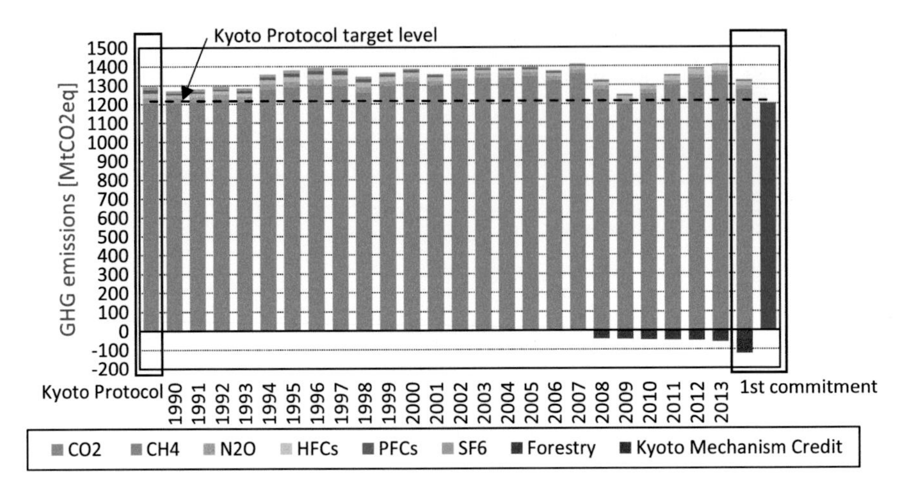

Fig. 4.2 GHG emissions from 1990 and Kyoto target (Data source: GIO 2014, 2015)

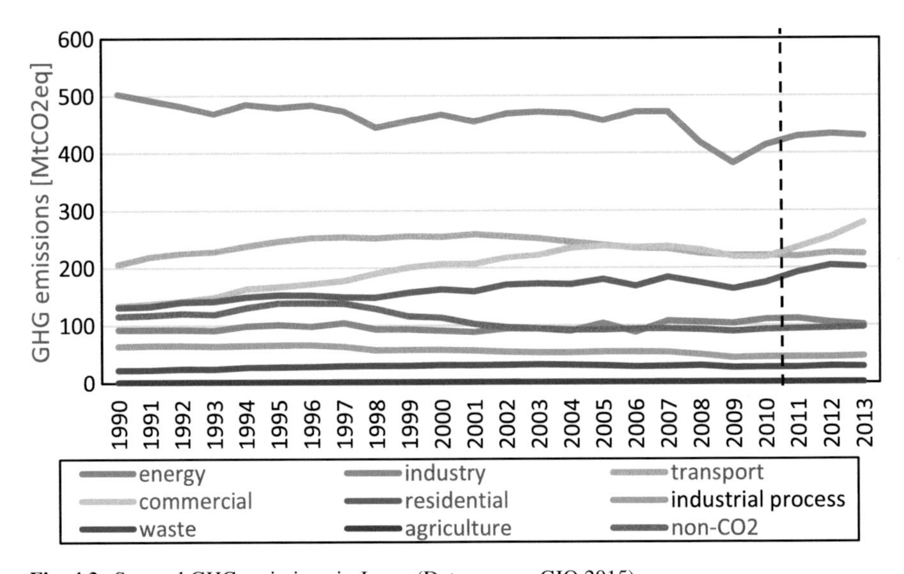

Fig. 4.3 Sectoral GHG emissions in Japan (Data source: GIO 2015)

Environment and Ministry of Economy, Trade and Industry. On April 30, the new GHG mitigation target in 2030 was announced by the government. The value is 1.042 $GtCO_2eq$, and this is 26.0 % reduction from the 2013 level. If the base year is set to 2005, the value of the reduction target becomes 25.4 %. The outline of the mitigation target in Japan in 2030 is introduced in Table 4.1.

The GHG mitigation targets in Japan have been changed when the government was changed. In 2008, in order to discuss the GHG mitigation target in 2020 at COP15 held in Copenhagen, the committee on the mid- and long-term target in Japan was organized at the Cabinet Secretariat. At that discussion, three types of

Table 4.1 Mitigation target in Japan in 2030

Unit: $MtCO_2eq$	Emissions in 2030	Emissions in 2005	Emissions in 2013
CO2 from energy combustions	927	1219	1235
Industry	401	457	429
Commercial	168	239	279
Residential	122	180	201
Transportation	163	240	225
Energy	73	104	101
Nonenergy CO_2	70.8	85.4	75.9
CH_4	31.6	39.0	36.0
N_2O	21.1	25.5	22.5
HFCs	21.6	12.7	31.8
PFCs	4.2	8.6	3.3
SF_6	2.7	5.1	2.2
NF_3	0.5	1.2	1.4

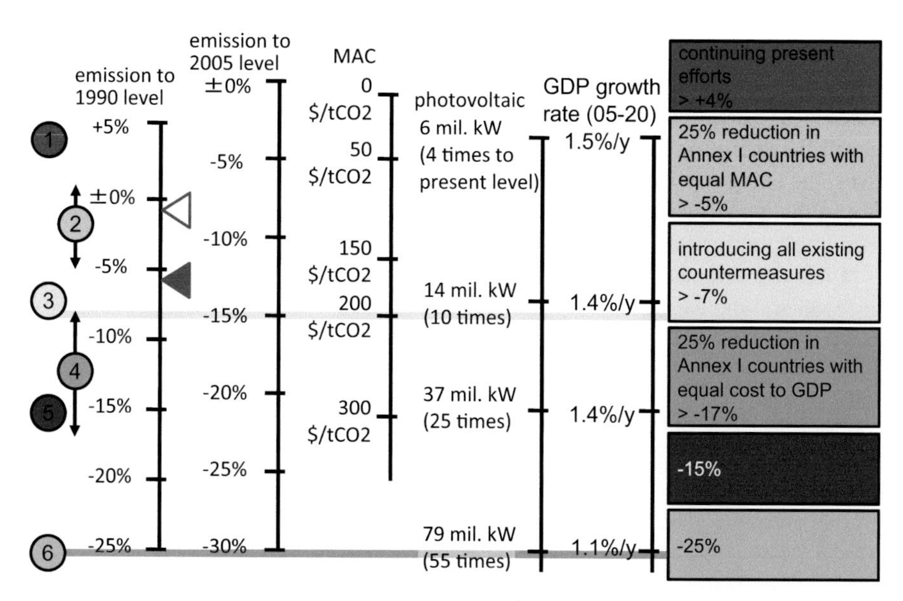

Fig. 4.4 Six options discussed in FY 2008 to FY 2009 (Note: The *red triangle* shows the mitigation target of all greenhouse gas emissions by the Kyoto Protocol. The *white triangle* shows the mitigation target of CO2 emissions from energy use)

models were utilized, global enduse model, Japan enduse model, and Japan economic model, and the proposed six options were quantified. Finally, 15 % reduction in 2020 compared with the 2005 level was selected as a target by former Prime Minister Mr. T. Aso. Figure 4.4 represents the contents of six options.

After the change of the government, the new Prime Minister, Mr. Y. Hatoyama, announced the 25 % reduction compared to 1990 at the United Nations Climate

Change Summit held in New York in 2009. In reaction to this new mitigation target, the road maps to achieve this 25 % reduction target were quantified at the Central Environment Council. But just after they completed the road maps to achieve the target, the Great East Japan Earthquake and the accident of Fukushima Daiichi Nuclear Power Plant of Tokyo Electric Power Company (TEPCO) happened on March 11, 2011. Due to the nuclear power accident, the GHG mitigation road map and the future energy mix should be reconsidered, because those at that time relied on nuclear power. In 2012, the Energy and Environment Council forecasted that the GHG emissions in 2020 would be 5–9 % reduction compared to 1990 level in the case of the prudent economic growth case.

After the next change of the government, the new GHG mitigation target was officially revised to be 3.8 % reduction compared to the 2005 level under the assumption of the no nuclear power at the COP19 of UNFCCC in Warszawa in 2013. But this target is equivalent to +3.1 % to the 1990 level. At that time, assessment using models was not implemented. Figure 4.5 shows the GHG emission forecasts in 2020 for the high economic growth case by using the AIM/Enduse [Japan] model executed in 2012. In this figure, the target of 2013 is represented by the red line. And Fig. 4.6 shows the relationship between the additional investment cost and saved energy costs. From this figure, the total saved energy costs for the lifetime of each equipment can exceed the total additional investment costs to install the energy-saving equipment. That is to say, if we have long-term perspectives, the energy-saving investment will bring the benefit to the economy in Japan. But in the actual world, the investment is decided from the short-term scope, and then, the introduction of the energy-saving technologies is difficult.

Table 4.2 summarizes the brief history of decision of GHG emission mitigation in Japan and the world.

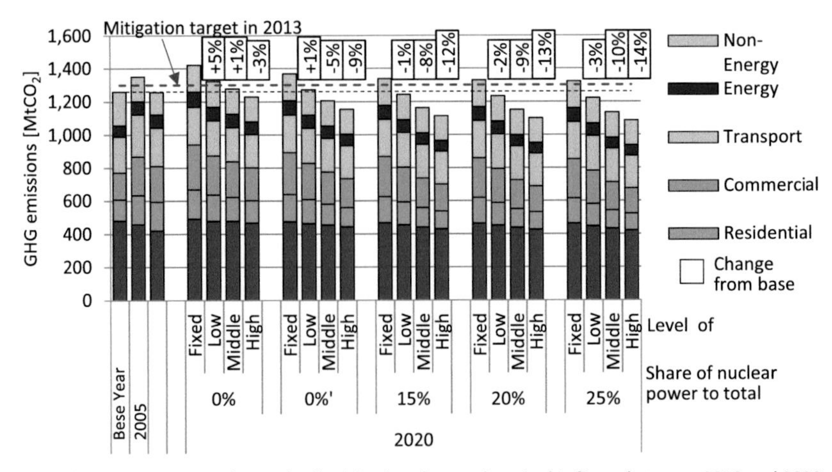

Share of nuclear power is set to be gradually shifted to the numbers in this figure between 2010 and 2030. "0%" is assumed to be 0% in 2020 and after

Fig. 4.5 GHG emissions in 2020 under the high economic growth case (as of June 2012)

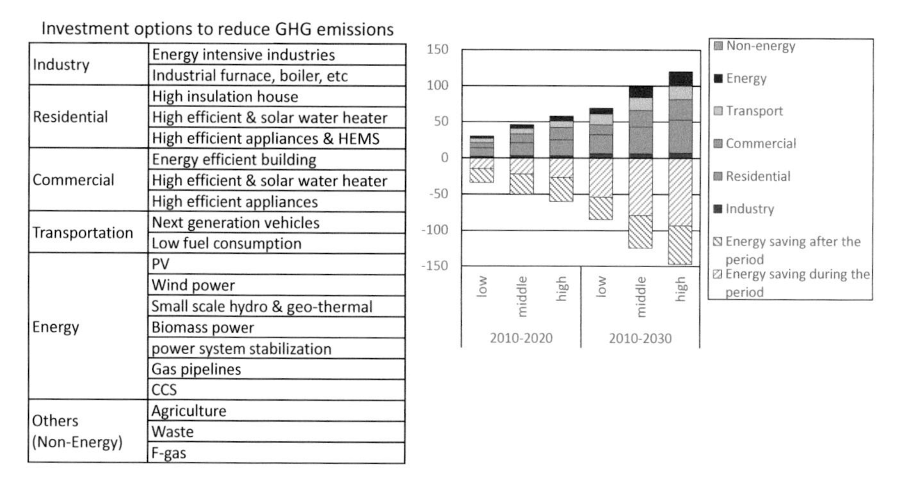
Investment options to reduce GHG emissions

Sector	Option
Industry	Energy intensive industries
	Industrial furnace, boiler, etc
Residential	High insulation house
	High efficient & solar water heater
	High efficient appliances & HEMS
Commercial	Energy efficient building
	High efficient & solar water heater
	High efficient appliances
Transportation	Next generation vehicles
	Low fuel consumption
Energy	PV
	Wind power
	Small scale hydro & geo-thermal
	Biomass power
	power system stabilization
	Gas pipelines
	CCS
Others (Non-Energy)	Agriculture
	Waste
	F-gas

Fig. 4.6 Necessary additional energy-saving investment and saved energy costs (as of June 2012)

4.3 2 °C Target and Mitigation in Japan in 2050

As shown in Fig. 4.1 from IPCC (2014), in order to achieve the 2 °C target, the global GHG emissions in 2050 would be 41–72 % compared to the 2010 level. Since the G8 Summit at Heiligendamm in 2007, the global leaders have shared the long-term vision that the GHG emissions in 2050 should be half compared to the present level. If GHG emission per capita is equal among the world in 2050, it becomes around 2 tCO_2. In the case of Japan, the present emission level is around 10 tCO_2. That is to say, the GHG emissions per capita should be 80 % reduction to achieve the 2 °C target.

In the 4th Environmental Basic Plan endorsed by the Cabinet in 2012, the long-term GHG emission mitigation target was made clear. The target in 2050 is 80 % reduction of GHG emissions. In the following sections, the possibility of 80 % reduction of GHG emissions is examined. The quantitative results are based on Kainuma et al. (2014) for the report of the Deep Decarbonization Pathways Project (DDPP). DDPP is organized by the Sustainable Development Solutions Network (SDSN) and the Institute for Sustainable Development and International Relations (IDDRI), and 15 countries including Japan and international organizations join this project. The interim report was opened at the Climate Summit in 2014. For more information, please see IDDRI and SDSN (2014).

Table 4.2 History of GHG emission mitigation in Japan and the world

Time	Japan	World
May 1992		Adoption of UNFCCC
December 1997	COP3 in Kyoto. Adoption of Kyoto Protocol During 1st commitment period (2008–2012), 6 % reduction compared with those in base year	
May 24, 2007	Invitation to "Cool Earth 50" by then Prime Minister Abe Halving global GHG by 2050 compared with present level	
June 8, 2007		G8 Summit at Heiligendamm: "we will consider seriously the decisions . . . which include at least a halving of global emissions by 2050"
November 2008–April 2009	Discussion of GHG mitigation target at Mid-term Target Committee, Cabinet Secretariat	
June 11, 2009	Then Prime Minister Aso announced GHG emission reduction target in 2020 15 % reduction of domestic emission compared to 2005 level	
September 2009	Address by then Prime Minister Hatoyama at the 64th session of the general assembly of the UN GHG emissions in 2020 will be reduced by 25 % compared to the 1990 level under fair and effective international framework	
October - December 2009	Discussion at task force meeting on climate change	
December 2009–March 2010	Discussion at the Investigative Commission on mid-/long-term road map to combat climate change: How to realize 25 % reduction by reconsidering assumptions and countermeasures	
January 2010	Based on "Copenhagen Accord," each country submitted mitigation target/action plan 25 % reduction, which is premised on the establishment of a fair and effective international framework in which all major economies participate and on agreement by those economies on ambitious targets	
April 2010–March 2011	Results from subcommittee on mid-/long-term road map, the Central Environment Council showed the society in Japan achieving 25 % GHG reduction in 2020 compared to 1990	
November - December 2010		Cancun Agreements at COP16: "Establish clear goals and a timely schedule for reducing GHG emissions over time to keep the global average temperature rise below two degrees"

(continued)

Table 4.2 (continued)

Time	Japan	World
2011.3.11	Great East Japan Earthquake and TEPCO Fukushima Daiichi Nuclear Power Plant accident	
July 2011-June 2012	Discussion at subcommittee on counter-measures/policies post 2013, Central Environment Council	
April 4 2012	The 4th Environmental Basic Plan	
	GHG emissions in 2050 will be reduced by 80 %	
September 2012	Options for energy and the environment by the Energy and Environment Council	
	In the low economic growth case, GHG emissions in 2020 will be reduced by 5–9 % compared to 1990 level. In the high economic growth case, 2–5 % reduction	
	GHG emissions in 2030 will be reduced by about 20 % compared to 1990 level	
	GHG emissions in 2050 will be reduced by 80 %	
November 2013	Then Environment Minister Ishihara announced the new emission target in Japan at COP19	
	3.8 % reduction in 2020 compared to the 2005 level	
March 2015	Deadline to submit the post-2020 target (INDC) by country	
April 30 2015	Government of Japan presented the proposal of GHG mitigation in 2030	
December 2015	COP21 in Paris. The post-2020 target will be decided?	

4.4 How to Achieve 80 % Reduction Target in Japan

As shown in the explanation of the 4th Environmental Basic Plan, the mitigation target in 2050 is set to be 80 % reduction. In order to assess the 2050 target, we utilized the AIM/Enduse to disaggregate Japan into 10 regions. The treated technologies are the same as the analyses for 2020 and 2030 using the national model, and the treatment of future is also the same, that is to say, recursive dynamics. On the other hand, the features of this model are that the local renewable energy potential can be reflected, interconnected line of electricity among the regions can be assessed, and so on.

Other important assumptions are nuclear power and carbon capture and storage (CCS). As regards the nuclear power, the following are the main assumptions:

- Lifetime is limited to 40 years for plants built in 1990 and 50 years for all other plants, and from 2013 to 2035, an additional three GW nuclear plants capacity is included based on the premises of New Policies Scenario of the World Energy Outlook 2013 (IEA 2013).
- Subject to these assumptions and maximum capacity factor of 70 % for all plants, electricity generation from nuclear plants represents about 50 TWh in 2050.

As for the CCS, the following assumptions are set:

- Geologic carbon storage potential.
- Complying with previous studies, CCS technologies are assumed to be available in 2025, and annual CO_2 storage volume is assumed to increase up to 200 $MtCO_2$/year in 2050.

In the illustrative scenario, the primary energy supply and demand in 2050 decreases to almost half compared to the 2010 level as shown in Fig. 4.7. In 2050, renewables and fossil fuels with CCS account for more than 50 % of primary energy supply. In the power sector, the nuclear power is assumed to be phased out gradually, and electricity generation from coal without CCS is entirely phased out by 2050. Renewable energy is developed over the mid- to long terms and reaches approximately 59 % of total electricity generation through large-scale deployments of solar PV and wind power. In addition, natural gas (equipped with CCS) is developed to ensure balancing of the network and reaches about a third of total electricity generation in 2050. Carbon intensity of electricity falls to nearly zero in 2050 as shown in Fig. 4.8. And also, in the final energy demand sector, the energy intensity will be improved, and electricity will be the dominant energy over the long term. As a result, the long-term GHG emission reduction target in 2050 will be reached as shown in Fig. 4.9.

The alternative pathways are also investigated using the same model. In the "without nuclear power scenario," an 80 % emission reduction in 2050 is still feasible. However, higher carbon intensity is experienced during the transition period. The impact of nuclear phaseout as compared to the illustrative scenario is

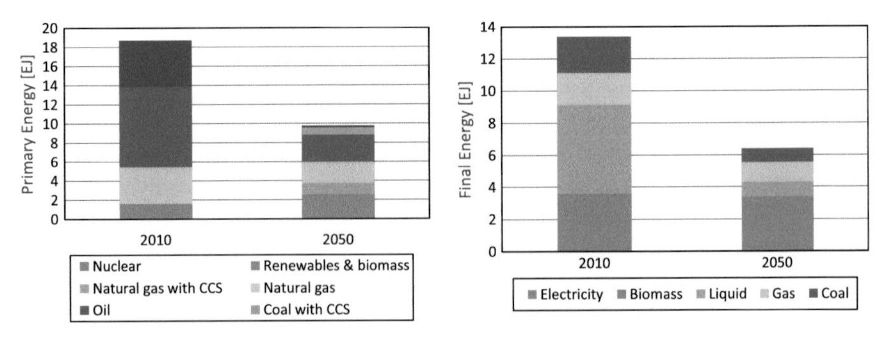

Fig. 4.7 Energy pathways by source

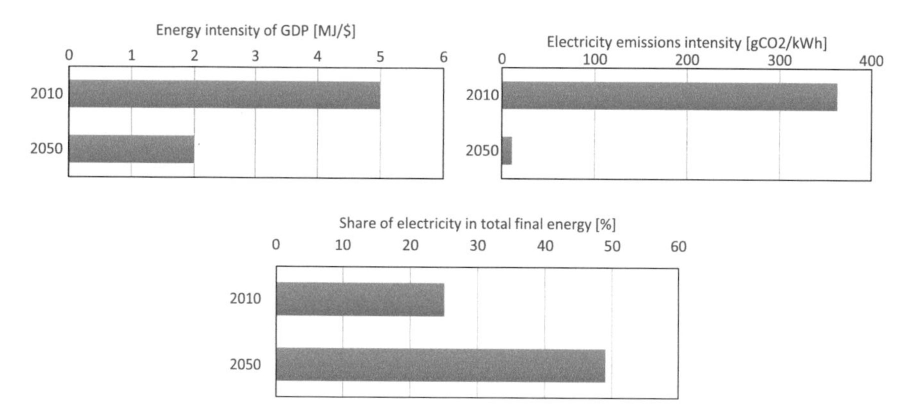

Fig. 4.8 Drivers of decarbonization

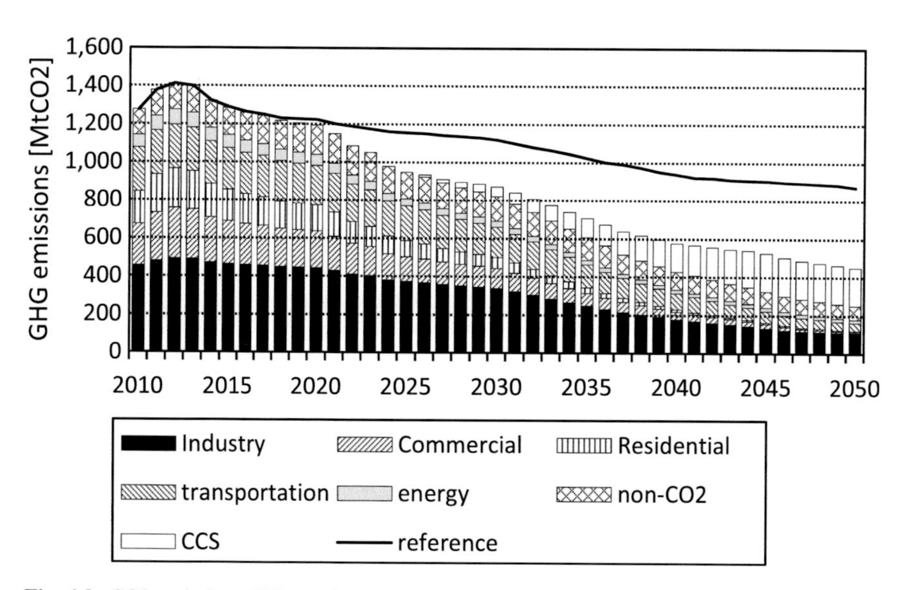

Fig. 4.9 CO2 emission of illustrative scenario from 2010 to 2050

relatively small in the long term, given the small share of nuclear energy in 2050 in any case. Another alternative scenario is "less CCS scenario," in which CO_2 storage volume is assumed to be limited to 100 $MtCO_2$/year. Achieving long-term emission reduction target proves to be still feasible with substantial increase of renewable energy, particularly solar PV and wind power. The share of renewable energy in electricity supply reaches approximately 85 % in 2050, and variable renewable energies account for about 63 % in electricity generation in 2050, hence imposing a further challenge for integration into the electricity system.

4.5 Conclusion

In this chapter, the historical change of the emission reduction target in Japan and the possibility of 80 % reduction in 2050 in Japan are represented. The feature of mitigation target in Japan is mainly based on the bottom-up approach. That is to say, the process stressed the feasibility of the target. On the other hand, the top-down decision is also requested for the ambitious reduction target.

From the previous estimations, even in Japan, there are still many reduction potentials. Toward the achievement of 2 °C target, taking actions with the long-term perspective becomes more important.

Acknowledgment The research in this chapter is supported by the Environment Research and Technology Development Fund by the Ministry of the Environment, Japan (2-1402).

References

GIO (2014) Japan's national greenhouse gas emissions in fiscal year 2012 (Final figures). http://www.nies.go.jp/whatsnew/2014/20140415/20140415001-e.pdf

GIO (2015) Japan's national greenhouse gas emissions in fiscal year 2013 (Final figures). http://www.nies.go.jp/whatsnew/2015/gaiyo-e.pdf

IDDRI and SDSN (2014) Pathways to deep decarbonization, 2014 report. http://unsdsn.org/wp-content/uploads/2014/09/DDPP_Digit_updated.pdf

IEA (2013) World energy outlook 2013. OECD/IEA, Paris

IPCC (2013) Climate change 2014: The physical science basis. Cambridge University Press, Cambridge and New York

IPCC (2014) Climate change 2014: mitigation. Cambridge University Press, Cambridge and New York

Kainuma M, Oshiro K, Hibino G, Masui T (2014) Japan chapter of pathways to deep decarbonization, 2014 report. http://unsdsn.org/wp-content/uploads/2014/09/DDPP_2014_report_Japan_chapter.pdf

Chapter 5
Potential of Low-Carbon Development in Vietnam, from Practices to Legal Framework

Nguyen Tung Lam

Abstract Vietnam is not in the category of mandatory reductions of greenhouse gas emissions. However, when implementing the mitigation of greenhouse gas (GHG) emissions, Vietnam has many opportunities to access financial resources, technology and capacity building from developed countries to develop in a sustainable manner toward a green economy with low carbon and contribute to efforts to reduce global GHG emissions. Vietnam should prioritize the sectors for GHG reduction while ensuring the objectives of economic growth, employment, and economic development.

The GHG emissions in the energy and agriculture, forestry and land use (AFOLU) sectors are two of the greatest GHG emissions. Policies to reduce GHG emissions also have negative and unintended effects. Therefore analysis and evaluation of the externalities of policies and measures to reduce GHG emissions are essential. The negative externalities are considered as indirect costs of GHG emission reduction measures, therefore they are important when considering the priority of GHG emission reduction.

International experience of accessing low-carbon development programs from low-carbon development research is a valuable reference for Vietnam. The Asia Pacific Integrated Model (AIM model) to project GHG emission scenarios helps to identify priority sectors that have high potential in reducing GHG and less effects on the development targets. Accordingly, for developing countries like Vietnam, when the budget is not abundant and also to serve multiple objectives of other urgent development, GHG emission reductions in selected priority sectors and actionable measures need less investment and other negative impacts on socio-economic development targets.

Research has contributed to the development of GHG emission reduction policies in Vietnam. It is considered as an important basis for construction, adjustment, and additional amendment of the legal system, mechanisms and policies to promote GHG emission reduction activities in industry and other sectors.

N.T. Lam (✉)
Institute of Strategy, Policy on Natural Resources and Environment, Ministry of Natural Resources and Environment, Vietnam, 479 Hoang Quoc Viet, Hanoi, Cau Giay, Vietnam
e-mail: ntlam@isponre.gov.vn

S. Nishioka (ed.), *Enabling Asia to Stabilise the Climate*,
DOI 10.1007/978-981-287-826-7_5

Keywords Vietnam • Low-carbon development • AIM model • GHG emission reduction • Mitigation

Key Message to Policy Makers

- In Vietnam, priority sectors should be identified to reduce investment for GHG reduction.
- Great potential exists in the energy, waste and AFOLU sectors in Vietnam.
- Both positive and negative effects of GHG emission reduction policies are identified.

5.1 Introduction

Low-carbon development considers reducing greenhouse gas emissions through reduced energy consumption by technological innovation and social attitudes. Some sectors have great potential to reduce carbon emissions such as energy, agriculture, industries, construction, and waste management. Recent research has shown that Vietnam has great potential to reduce GHG emissions in the energy sector or agriculture. However, to implement a low-carbon development strategy requires large financial capacity, high-tech capabilities and appropriate supporting policies. Besides, improper awareness about the benefits of implementing a low-carbon development strategy, for example, like expensive investment but no immediate economic returns, would be a challenge to successful low-carbon development implementation in Vietnam.

As a developing country, Vietnam has no obligation to reduce emissions in the present, but with implementation of the action program of voluntary reductions of GHG emissions, Vietnam has many opportunities to receive support from other developed countries to develop its economies toward low carbon, and also has an opportunity to contribute to global GHG emission reduction efforts. In a developing country like Vietnam, a policy to develop low carbon will benefit all aspects: reducing energy consumption, increasing energy efficiency, saving natural resources, technological modernization, increased levels of economic value added, and elimination of environmental pollution. This is an opportunity that Vietnam can take advantage of in the future.

Recognizing the importance of implementation of practical actions to respond to climate change, the Government of Vietnam has issued many related legal documents. From 2006 to 2010, the Government of Vietnam adopted many important policies such as the National Strategy for Environmental Protection, the National Target Program for Energy Saving and Efficiency, the National Target Program to Respond to Climate Change, the National Strategy for Solid Waste Management, the National Green Growth Strategy, etc., and promoted the economy toward low carbon. This is an important legal basis for the implementation of sustainable development policy in practice, toward a low-carbon economy in Vietnam. However, to effectively implement the policy, it requires the coordination and cooperation of many agencies and departments from the central to local levels. In particular, the successful experience throughout the world has demonstrated that policy measures toward strategic development of low carbon will provide practical effects if they are confirmed in terms of technology, trade, and economics; socially accepted; and put into a legal framework for implementation.

To be able to implement development toward low carbon effectively, it should be determined what areas of the economy will play a key role in cutting emissions, the level of reductions, and a roadmap of implementation reduction measures in selected economic sectors. The formulation and promulgation of a low-carbon development policy should also consider their potential impacts on the economy such as creating jobs, changes in national income, changing industry structure, economic scale investment requirements and necessary resources to carry out measures for each respective sector. In some countries around the world, especially in developed countries that have committed to the reduction of GHG emissions, growth toward low-carbon emissions is considered as an integral part of a national strategy on climate change.[1] Thus quantitative research on low-carbon development arises as an essential need to provide a scientific basis for management decisions about the goals, schedule and reasonable solutions for strategic planning for economic development in the direction of decreasing GHG emissions. Recently, low-carbon development has received the attention of developing countries where the demand for fossil fuel has been increasing to meet the economic growth in the context of energy efficiency, which is still low. The study of low-carbon development has begun to be deployed in a number of countries with adjustments to suit their socio-economic conditions.

Although still relatively new in Vietnam, the recent problems of low-carbon development have received increasing attention of governments, international donors and agencies. In the implementation process of responsibility to participate in international exchange on climate change, the Ministry of Natural Resources and Environment has proposed to the Government policies and strategies for promoting low-carbon growth and a roadmap to reduce GHG emissions in Vietnam. The Prime Minister approved the National Green Growth Strategy, in which the economic development model toward low-carbon emissions is mentioned as important content of the strategy.[2]

In countries with emission reduction obligations, their policy will be anchored at the cutting rate that was committed to. Thus the trade-off between economic growth targets and the level of GHG emission reduction makes the cutting costs in

---

[1] *Low-Emission Development Strategies (LEDS), Technical, Institutional and Policy Lessons. Clapp et al. OECD (*2010*). Available at* http://www.oecd-ilibrary.org/environment/low-emission-development-strategies-leds_5k451mzrnt37-en?crawler=true

[2] *Decision 1393/QĐ-TTg dated 25/09/2012. "Chiến lược Tăng trưởng xanh quốc gia"* (2012, *in Vietnamese). Available at:* http://vanban.chinhphu.vn/portal/page/portal/chinhphu/hethongvanban?_page=1&class_id=2&document_id=163886&mode=detail

developed countries very high. For Vietnam, the reduction of GHG emissions should be voluntary so it may optionally give GHG reduction targets. Thus the problem for Vietnam is not how great the trade-offs are but which reduction plans would be preferred alternatives. Identification of the objectives, contents and methods in each sector's cuts and the transformation roadmap to achieve the level of emission reductions cannot follow idealistic sentiments; they must be based on scientific methods of calculation taking into account the specific conditions of the national economic potential of the country, the context of international relations and the requirements of meeting the government's socio-economic development targets.

This paper presents an overview of the recent situation of GHG emissions in Vietnam, as well as a discussion on how the country should select priorities in selecting emission alternatives. It has four main parts; the first one introduces the context of socio-economic development in which GHG emission reduction has been considered as a commitment of the government to the international community for its contribution to the global GHG reduction efforts. The second part presents in detail the main GHG emissions from different sectors and their projections in the years to come. The next part looks at the priority in selecting the alternatives for GHG emissions that are required to help the country to be balanced with its socio-economic development targets. The final part focuses on the impacts of these reduction policies on the country's development with suggestions to reduce unintended effects.

5.2 GHG Emissions in Vietnam

5.2.1 The Total Amount and Level of Greenhouse Gas Emissions in Vietnam

Vietnam signed the United Nations Framework Convention on Climate Change (UNFCCC) in 1992 and ratified this Convention in 1994. Vietnam also signed the Kyoto Protocol (KP) in 1998 and officially approved it in 2002. Under the Kyoto Protocol, Vietnam is not in the group of countries that have a responsibility to reduce greenhouse gases.

Regarding actual GHG emissions, Vietnam is a country with low GHG emissions in the world. The emissions in 2000 were only about 150 million tons, out of 34,000 million tons of CO_2 equivalent emissions worldwide (that is equivalent to approximately 0.44 %). However, it should be recognized that the emission rate per capita in Vietnam, although lower than those in China, Korea and Thailand, is growing faster than the rates in those nations. Specifically, emissions have increased by nearly 6 times, from 0.3 tons CO_2/person in 1990 to 1.71 tons CO_2/person in 2010, while China's emissions increased by 3 times, Korea's increased by 2.5 times and Thailand's increased by 2 times (Figs. 5.1 and 5.2).

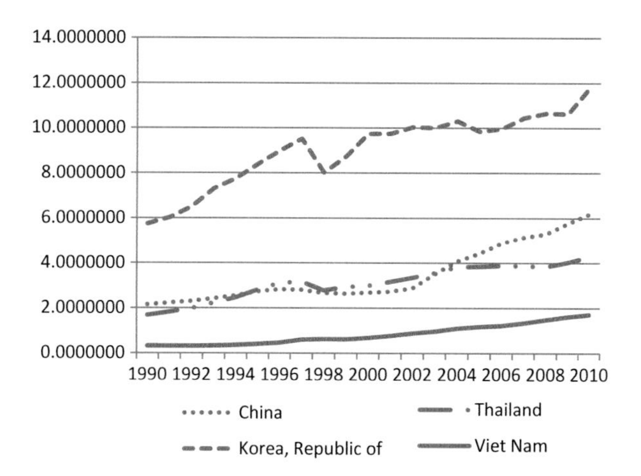

Fig. 5.1 GHG emissions per capita during the period of 1990–2010 (Source: Compiled from UN data at http://data.un.org/Data.aspx?d=MDG&f=seriesRowID:751)

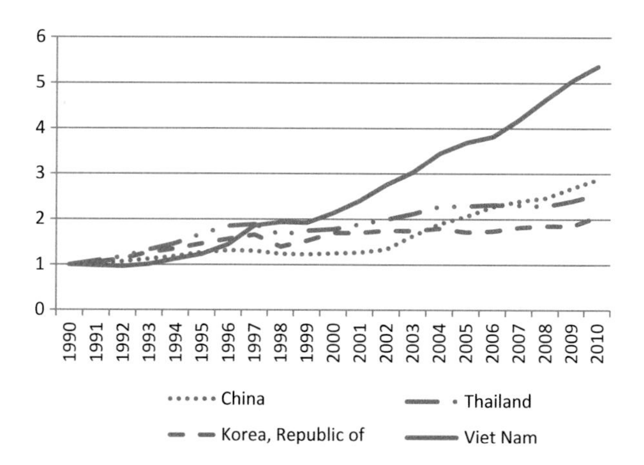

Fig. 5.2 The growing rate of GHG emissions per capita in Vietnam compared with some other countries (Source: http://data.un.org/Data.aspx?d=MDG&f=seriesRowID:751)

In the recent period (2001–2011), before the trend of economic development with a relatively high rate (average growth of 6–8 %), the increase in population led to the amount of Vietnam's greenhouse gases increasing. As expected, due to the economic development needs in the coming years, the amount of GHG emissions in Vietnam may be increased if there is not timely implementation of measures to reduce GHG emissions caused by economic development activities.

Vietnam conducted national GHG inventories for the years 1994, 2000 and 2005. This was to meet the country's commitments under the UNFCCC, also aiming to develop a database to support the formulation of policies related to climate change and greenhouse gases. The inventories therefore covered most sectors' GHG emissions in Vietnam.

All inventories were calculated using the Intergovernmental Panel on Climate Change (IPCC)'s 1996 guidelines for non-Annex I nations (Revised 1996 IPCC Guidelines for National Greenhouse Gas Inventories). The inventories for 2000 and

Table 5.1 GHG inventories in Vietnam

Year	1994[a]		2000[b]		2005[c]	
Sector	CO_2 t/đ	%	CO_2 t/đ	%	CO_2 t/đ	%
Energy	25,637.09	24.7	52,773.46	35.0	101,934.90	56.0
Industries	3,807.19	3.7	10,005.72	6.6	14,590.82	8.0
AFOLU						
Agriculture	52,450.00	50.5	65,090.65	43.1	83,828.40	46.1
LULUCF	19,380.00	18.7	15,104.72	10.0	−27,020	−14.8
Waste	2,565.02	2.4	7,925.18	5.3	8,643.41	4.7
Total	**103,839.30**	**100.0**	**150,899.73**	**100.0**	**181,977.53**	**100.0**

Source: Compiled from Vietnam Second Communication Report, Ministry of Natural Resources and Environment (2010), and Interim Report of Inventory Capacity Building Project. JICA (2014)
[a]Second Communication Report, MONRE 2010
[b]Second Communication Report, MONRE 2010
[c]Interim Report of Inventory Capacity Building Project. JICA (as of 6/2014)

2005 were combined with the Good Practice Guidance versions from the IPCC for 2000 and 2003 in a number of areas.

The GHG inventory was conducted in economic sectors that have high emissions, including energy, industrial processes, agriculture, and land use–land use change–forestry (LULUCF), and waste sectors. GHG inventories cover three major categories including CO_2, CH_4 and N_2O.

5.2.2 *Structure and Trends in Greenhouse Gas Emissions in Vietnam*

A summary of the national GHG inventories in 1994, 2000 and 2005 is given in Table 5.1. The data in the table are the total amounts of GHG emissions in the base year and are converted into CO_2 equivalents. Figure 5.3 shows the trends in GHG emissions from different sectors in the inventory periods.

Excluding the absorption from LULUCF, the volumes of GHG emissions from activities in the industrial, energy, agriculture and waste management sectors also tended to increase, but by different amounts. Among those, emissions from the energy sector have been the fastest rising trend. The change in the structure of GHG emissions as a result of the third inventory excluding the LULUCF sector is represented and trends are shown in Fig. 5.4.

5.2.3 *Trends in Emissions from Different Sectors*

Emissions from industrial processes and waste account for a small proportion of GHG emissions in Vietnam. With the economic development trend toward green

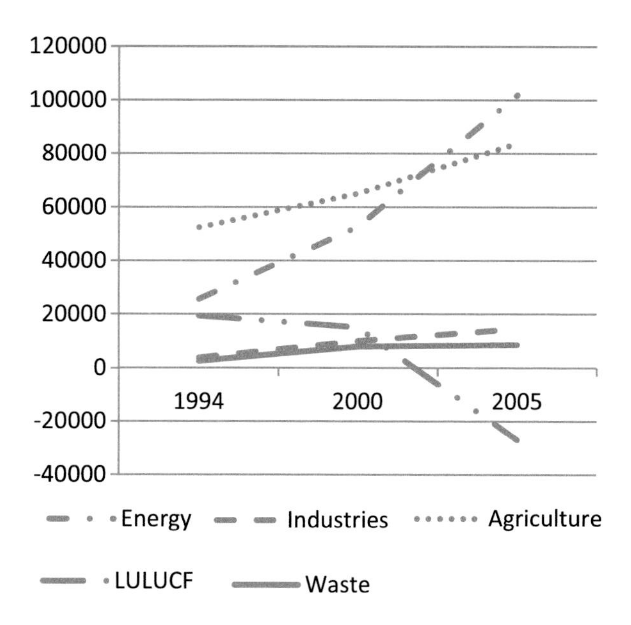

Fig. 5.3 Total GHG emissions from different sectors in the inventory periods (×1000 tCO_2e) (Source: Compiled from Vietnam Second Communication Report 2010, and Interim Report of Inventory Capacity Building Project. JICA 2014)

growth and low-carbon development, the industries that have high potential for emissions, such as cement, steel, and chemicals, will not likely be developed at high speed to create a larger proportion of the total emissions, while emissions from the waste sector will remain at the same level. Urban development will require accompanied waste minimization and management solutions will reduce the environmental pollution and GHG emissions.

The sectors that currently have the largest proportion of emissions are agriculture and energy. However, emissions from energy will tend to increase rapidly in the coming years in terms of total volume (Fig. 5.3) as well as the proportion of the emission structure (Fig. 5.4). As is likely in most other countries, the energy sector will account for the largest emissions in the economic structure of the country in the years to come.

In the previous year, emissions from the agricultural sector accounted for over 50 % of the components of Vietnam's GHG emissions and emissions of CO_2 and CH_4 (mainly from the energy sector) accounted for approximately 50 %. However, the trend in emissions from energy will increase and serve as the main source of emissions in Vietnam in the coming years, and CO_2 will be the main GHG emissions in Vietnam, beyond emissions of CH_4 from agriculture and waste.

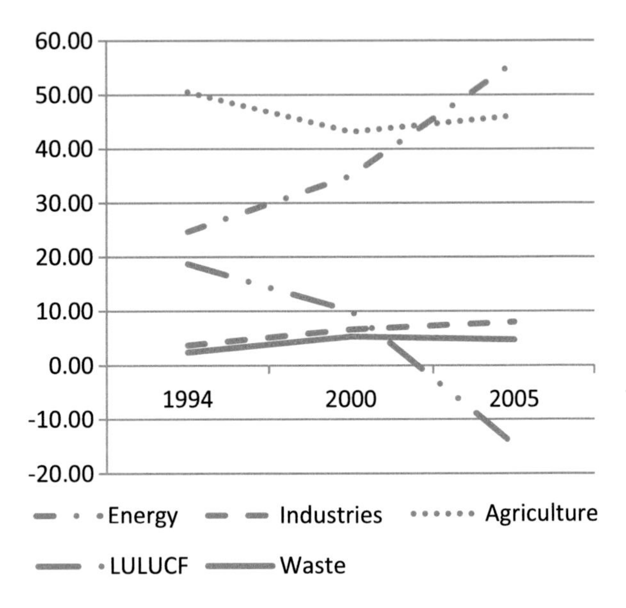

Fig. 5.4 Trends in the proportions of GHG emissions from different sectors (Source: Vietnam Second Communication Report 2010, and Interim Report of Inventory Capacity Building Project. JICA 2014)

5.2.4 Greenhouse Gas Emissions from Different Sectors

5.2.4.1 The Energy Sector

The energy sector has had important implications for the process of sustainable development of the national economy and people's lives in recent years. Vietnam's energy sector has contributed significantly to the country's development, industrial growth and exports. The total primary energy consumption of Vietnam for the period of 2000–2009 showed an average increase of 6.54 %/year and reached 57 million toes (tons of oil equivalents). In 2009, the average coal consumption increased 12.12 %/year, fuel 8.74 %/year, gas 22.53 %/year, and power 14.33 %/ year, reaching 74.23 billion kWh.

The total final energy consumption in 2000 was 26.28 million toes, which increased to 40.75 million toes in 2007, during which the proportion of coal consumption increased from 12.3 to 14.9 %, gasoline consumption increased from 26.3 to 34.4 %, gas consumption increased from 0.1 to 1.3 % and electricity consumption increased from 7 to 12.9 %. Regarding the structure of energy consumption by different sectors, this has changed; in 2000, 30.6 % of energy consumption was in industry, 14.7 % was in transport, 1.5 % was in agriculture, 48.8 % was in the residential sector, and 4.4 % was in commercial services. By 2007, the proportion in industry rose by 34.3 %, agriculture increased by 1.6 %, transportation increased by 21.2 %, the civil sector proportion dropped 39.1 %, and commercial services.[3]

[3] *Vietnam Second Communication Report* (2010), *and Interim Report of Inventory Capacity Building Project. JICA* (2014).

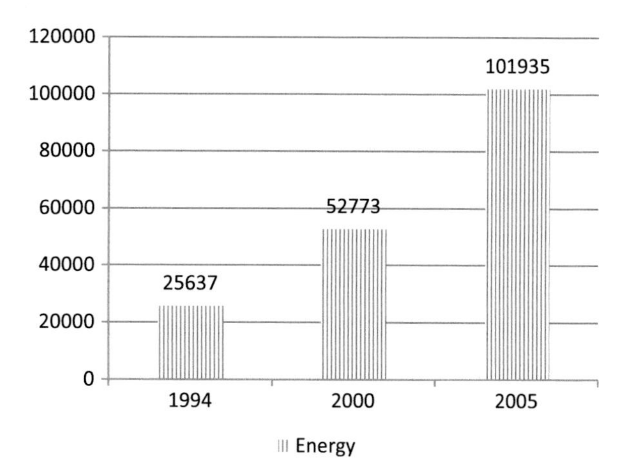

Fig. 5.5 Total GHG emissions from the energy sector in the inventory periods ($\times 1000$ tCO_2 e) (Source: Vietnam Second Communication Report 2010, and Interim Report of Inventory Capacity Building Project. JICA 2014)

These figures suggest that in addition to contributing to the country's economic development, the increasing exploitation and use of fossil fuels (coal, oil, gas) for energy have increased GHG emissions. In the energy sector, GHG emissions come from fuel combustion, mining activities and transportation. The main types of emission inventories in the energy sector include (1) GHG emissions from fuel combustion and (2) emissions from GHG emissions. GHG emissions from fuel combustion are divided into sub-sectors: electricity, industry and construction, transportation, trade/services, civil, agriculture/forestry/fisheries, and other. Emissions from GHG emissions are due mainly to coal, oil, gas and gas leaks.

GHG inventory results in the energy sector for the years 1994, 2000 and 2005 are shown in Fig. 5.5.

In total, emissions calculated over the time inventory of GHG emissions from the combustion of fuel account for about 85–90 %, and the rest is due to leakage from the fuel extraction process (coal, oil and gas), storage and transport of fuel.

5.2.4.2 Industrial Processes

The position of industries is increasingly being confirmed in the national economy; the industries are increasingly rich and diverse, ensuring the supply of products and raw materials essential for both consumption and production.[4] Export values of industrial production (in 1994 constant prices) in 2010 were estimated at 795.1 trillion VND, 4.0 times more than in 2000. In the 10 years from 2001 to 2010 the average annual increase was 14.9 %, while the state sector increased 2.1 times, an average annual increase of 7.8 %; the non-state area increased 6.5 times, an average annual increases of 20.5 %; regional foreign investment increased more than 4.7 times, an average annual increase of 16.7 %.

[4] Ministry of Industry and Trade 2013.

A number of important industrial products for production and consumption have reached a relative high with population growth. The output of coal in 2010 reached 44.0 million tons, 3.8 times the output in 2000, an average annual increase of 13.7 % over the 10 years from 2001 to 2010; 7.9 million tons of rolled steel were produced, a 3.5-fold increase, with an average annual increase of 17.5 %; 55.8 million tons of cement, a 3.8-fold increase, 15.4 %/year; 2.6 million tons of chemical fertilizers, a 2.1-fold increase, 7.8 %/year; 1887.1 thousand tons of paperboard, a 4.6-fold increase, 16.5 %/year; 1.2 billion m^2 of silk, a 3.4-fold increase, up to 13 %/year; 436.3 million boxes of condensed milk, a 1.9-fold increase, 6.7 %/year; 2.4 billion liters of beer, a 3.1-fold increase, 11.8 %/year; and 91.6 billion kwh of electricity were generated, a 3.4-fold increase, 13.1 %/year.

In addition to these results, the development of industries has exposed many shortcomings: low added value and a downward trend, with investment inefficiency and low technology levels.

GHG emissions from industrial processes are not the form of emissions related to energy use in the industrial processes. The emissions considered are those generated by the interaction of the physics and the chemistry of the material during material processing. Heavy and chemical industries in Vietnam so far are only at modest levels, so their GHG emissions are only considered as secondary sources. This is also consistent with the general trend throughout the world.

The GHG inventory for the period between 2000 and 2005 estimated emissions for different types of manufacturing industry, including cement, lime production, ammonium production, carbide production, and iron and steel production. In the first GHG inventory for 1994, the emissions also included the production of paper, alcohol and processed foods. However, the proportion of emissions from these activities is very small, so they are not included in the latter GHG inventories.

GHG inventory results for industry are shown in Fig. 5.6.

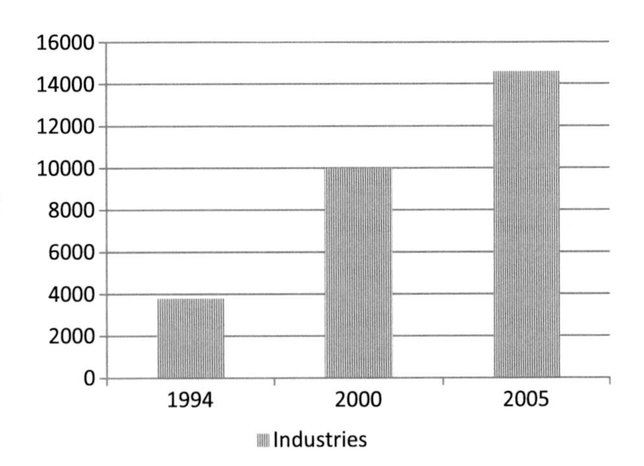

Fig. 5.6 Total GHG emissions in industrial processes in the inventory periods ($\times 1000$ tCO_2 e) (Source: Vietnam Second Communication Report 2010, and Interim Report of Inventory Capacity Building Project. JICA 2014)

5.2.4.3 The AFOLU Sector

Agriculture

The agricultural sector in the period of 2001–2010 saw steady growth, providing products with improved quality to better meet the needs of production, domestic consumption and export. The value of agriculture, forestry and fisheries (in 1994 constant prices) in 2010 was estimated at 232.7 trillion VND, up 66.4 % compared with the year 2000. The structures of agriculture, forestry and fisheries have transferred toward reducing the proportions of agriculture and forestry, with fisheries increasing in density. In 2000, the value of agricultural production (at current prices) accounted for 79 % of the total output value of agriculture, forestry and fisheries, and forestry and fishing accounted for 4.7 % and 16.3 %, respectively; by 2010 the proportions were 76.3 %, 2.6 % and 21.1 %, respectively.[5]

In addition to these achievements, the agricultural sector also has some drawbacks such as low-quality products and low value added. Development that has mainly focused on exploiting the potential of land, resources and labor rather than investment in cultivation and processing technologies has led to low-quality products. These inadequacies also lead to negative impacts of agriculture on the environment and ecology, which must be considered in terms of the increasing emissions of greenhouse gases from the types of agricultural activity.

The GHG emissions in agriculture come mainly from activities such as rice farming, raising livestock, emissions from arable land, and burning of agricultural products. The GHG emissions from agricultural activities are CH_4 and N_2O. The agricultural activities considered in calculation of the GHG emission inventory include enteric fermentation, livestock manure management, rice cultivation, agricultural soils, and field burning of agricultural residues. Among the agricultural activities, water rice cultivation account for most GHG emissions (45–60 %), followed by emissions from agricultural soils, enteric fermentation from cattle, and emissions from cattle manure. Other activities make up only a small proportion of emissions. Aggregate emissions from the agricultural sector are presented in Fig. 5.7

According to the 1994 GHG inventory, GHG emissions from the agricultural sector were 52.45 million tonnes of CO_2 equivalents, accounting for 50.50 % of total GHG emissions in the country. By the year 2000 this had changed to 65.09 million tonnes of CO_2 equivalents, accounting for 43.10 % of the total national GHG emissions (including emissions from rice cultivation, which accounted for 57.50 %; 21.85 % came from agricultural soils; 11.88 % came from enteric fermentation, and the rest came from manure management, and field burning of agricultural residues). According to data from the GHG inventory in 2005, GHG emissions from the agricultural sector were 83.828 million tons of CO_2 equivalents, accounting for 46.10 % of total GHG emissions in the country (including emissions

[5] General Statistic Office (2014).

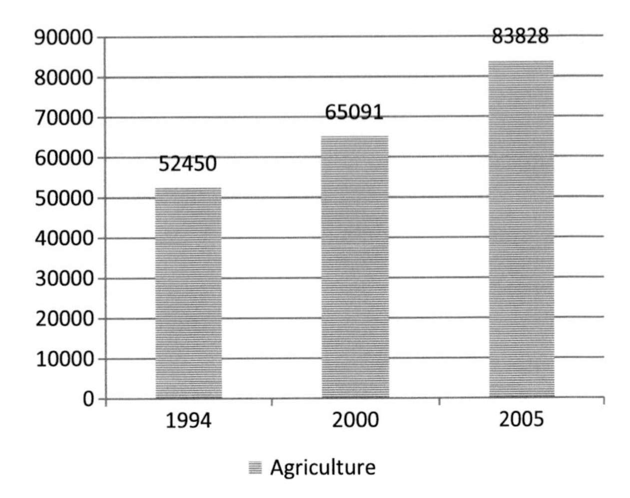

Fig. 5.7 Total GHG emissions from the agriculture sector (×1000 tCO_2 e) (Source: Vietnam Second Communication Report 2010, and Interim Report of Inventory Capacity Building Project. JICA 2014)

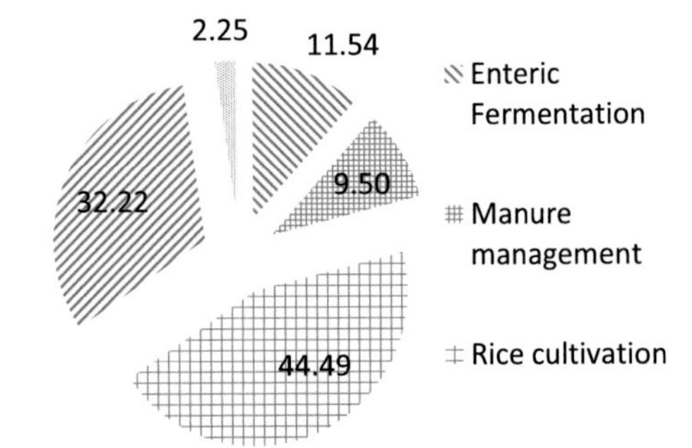

Fig. 5.8 Proportion of GHG emissions from the agriculture sector in 2005 (Source: Interim Report of Inventory Capacity Building Project. JICA 2014)

from rice cultivation, accounting for 44.49 %; 32.22 % came from agricultural soils, 11.54 % came from enteric fermentation, and the rest came from manure management and field burning of agricultural residues) (Fig. 5.8).

Land Use, Land Use Change and Forestry

Forests have roles as both emission sources and GHG sinks. Activities such as land use change and forest exploitation are the source of CO_2 emissions. Meanwhile, the activities of forest protection, reforestation and afforestation are sinks. Forestry, land use and land use change are areas of great potential GHG absorption through the reservoir of carbon from forests, soil, and vegetation if they are well managed, protected and appropriately exploited. The estimation of GHG emissions and absorption in this field focuses on the following main groups of activities: changes in the reserve forest area and biomass in natural forests and plantations; conversion of land use from forest land to other land; abandoned land management; and emission and absorption of CO_2 from the soil. Change of land use often causes

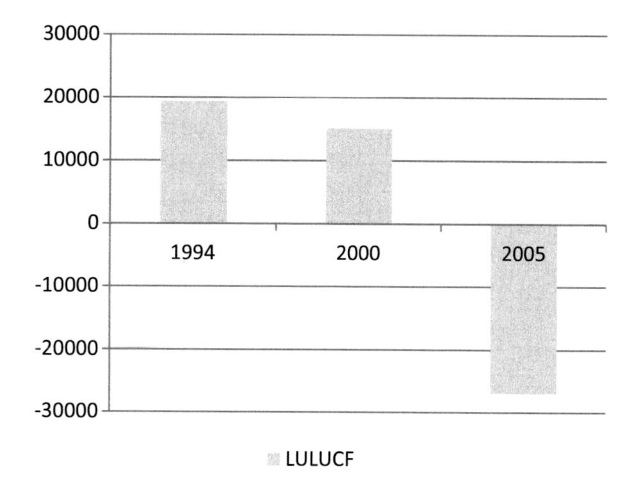

Fig. 5.9 Total GHG emissions from the LULUCF sector (×1000 tCO_2 e) (Source: Vietnam Second Communication Report 2010, and Interim Report of Inventory Capacity Building Project. JICA 2014)

more CO_2 emissions, while change in forest area (in term of increases) often leads to increased levels of CO_2 absorption. The inventory results for this area over the inventory period are given in Fig. 5.9.

According to the 1994 GHG inventory, the amount of GHG emissions in the field of forestry and land use change was 19.38 million tons of CO_2 equivalents, accounting for 18.70 % of total GHG emissions in the country. The GHG inventory in 2000 estimated that the emissions in the forestry and land use change sectors was 15.10 million tons of CO_2 equivalents, accounting for 10 % of the total national GHG emissions. The corresponding figure in the 2005 GHG inventory was −27.02 million tons of CO_2 equivalents, representing −14.8 % of total GHG emissions in the country. Thus, the LULUCF sector has become a major greenhouse gas sink in Vietnam. There is no satisfactory explanation for this sudden change in the calculated results. These issues will also need to be discussed for clarification. But it is clear that the forestry, land use and land use change sector in Vietnam has great potential for GHG absorption through the reservoir of carbon from forests, soil, and vegetation, if they are well managed, protected, and appropriately and sustainably exploited and used.

Waste Management

GHG emissions from the waste management sector are calculated for collected and disposed municipal solid wastes and GHG emissions from domestic sewage and industrial wastewater. It is estimated that every year about 15 million tons of solid waste is discharged from various sources, of which over 80 % are from urban areas, and the rest is industrial waste. However, only part of this waste is collected and processed; the data show that the proportions are over 70 % in urban areas and more than 20 % in rural areas.

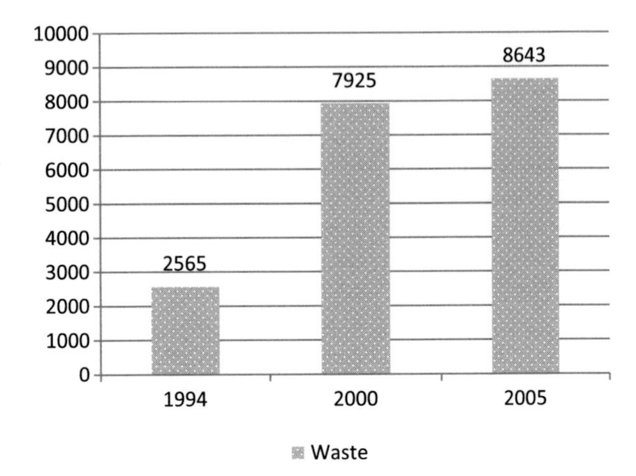

Fig. 5.10 Total GHG emissions from the waste sector (×1000 tCO_2 e) (Source: Vietnam Second Communication Report 2010, and Interim Report of Inventory Capacity Building Project. JICA 2014)

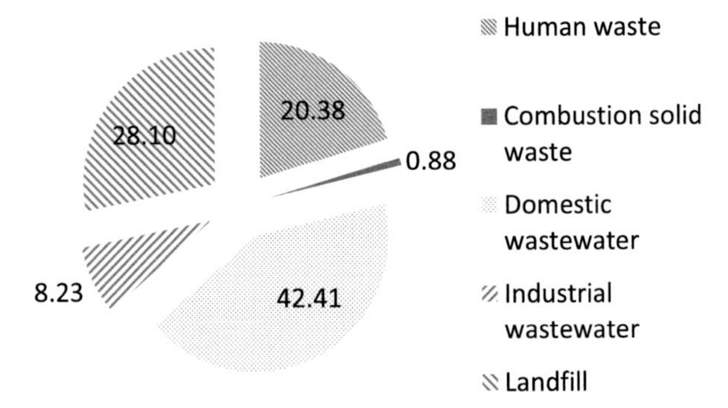

Fig. 5.11 Proportion of GHG emissions from the waste sector in 2005 (Source: Vietnam Second Communication Report 2010, and Interim Report of Inventory Capacity Building Project. JICA 2014)

The calculation of GHG emissions from the waste sector in GHG inventories for Vietnam focuses on the main sources of emissions including CH_4 emissions from solid waste landfills; CH_4 emissions from industrial wastewater and domestic sewage; N_2O emissions from domestic sewage sludge; and CO_2 and N_2O emissions from the incineration of waste. The inventory results are given in Figs. 5.10 and 5.11

In the waste sector, emissions from domestic wastewater are the largest, which are estimated about 3.4 million tons, accounting for about 42 %; the emissions from landfill waste are 2.3 million tons, accounting for 28 %, and the emissions from human waste are approximately 1.69 million tons. Emissions from combustion of solid waste are not high, only about 0.9 % by incineration operations, as this technology is not popular. To reduce GHG emissions from the waste sector requires a focus on the areas of domestic sewage and solid waste landfill.

The data on the status of GHG emissions in Vietnam have shown that two areas that have high levels of emissions are energy and agriculture, while two other sectors, industrial processes and waste, have much lower levels. Depending on the characteristics and properties of GHG emissions in four areas, it shows that to reduce GHG emissions from the energy sector and industrial processes requires large investments with wide impacts on socio-economic aspects; while the potential to reduce GHG emissions from the AFOLU sector and waste management may require lower funding with fewer impacts on the economics. In particular, the LULUCF sector also has high potential for GHG absorption. However, to get a clearer view of the potential to reduce GHG emissions, it is necessary to analyze and assess the opportunities and challenges for reducing GHG emissions in Vietnam.

5.3 Identification of External Impacts of GHG Emission Reduction Policies

5.3.1 *Externalities of Greenhouse Gas Emission Policies*

In the context of climate change impacts, the debate about the achievements and negative impacts of policies to reduce GHGs is becoming more and more popular. Recently, this problem has attracted a more extensive and more detailed focus on the benefits and costs of the externalities of the policy options and mitigation plans. Basically, these externalities can be understood as policies and plans to reduce greenhouse gases that could, in some way, cause a positive or negative impact on the economy, public health, ecosystems, etc. And, in this case, the effects can be monetized; they should be subtracted or added to the social cost of emission reduction policies. The positive externalities can be created through minimizing damage to the environment and the health of the pollutants. Conversely, these policies can also create negative externalities for public health and the environment, for example, in the field of energy, the increased use of diesel fuel can reduce GHG emissions, but will increase the risks to environmental and human health. Generally, these externalities have not so far been studied and evaluated fully, and so rarely have been quantified in a systematic way and integrated into the emission reduction policies. Failure of consideration and evaluation of the impacts of externalities may affect the choice of policies to reduce emissions. The externality, accordingly, should be considered as one of the indicators to identify the priorities for policies to reduce GHG.

5.3.2 *The Impact of Macroeconomics*

Policy impacts on the energy sector such as fossil fuel price rises, or policies imposed on the industrial sector such as rising commodity prices related to GHG emissions, can help reduce emissions as well as the risks of climate change in the long term; conversely, however, they also reduce economic activity in many forms.

Although these effects on the economy's growth may not be large in the long term, they need to be considered. The policies on reducing GHG emissions that impact on economic activities can be generalized as follows:

- The shift in production, investment and labor from industries related to energy production based on carbon, or products and services that use a lot of energy, to industries using alternative energy sources and consuming less energy;
- Reduced productivity of capital and labor in accordance with the cheap energy available;
- Reduced household incomes, with a reduction in domestic reserves;
- Lack of encouragement for investment due to increasing capital costs of production processes using a lot of energy;
- Reduced amount of net income from abroad (decreased productivity and increased cost of production capital), making the domestic market become less attractive to foreign investors;
- Deterioration of total labor supplies due to increases in the cost of consumer goods and reductions in the real wages of workers.

The GHG emission reduction policies may affect GDP growth through investment mechanisms. For example, high taxes on production that has a high level of GHG emissions will cause increases in production costs, thereby reducing investment and leading to a decline in product supplies and real wages. Accordingly, the consumption by people will fall and, as a result, reduce GDP. At the same time, lower wages can reduce workers' choices for employment that is unpaid or is not reflected in GDP, such as parenting, employment at home or entertaining.

5.3.3 The Problems of Hunger Eradication and Poverty Reduction

Solutions to reduce GHG emissions can cause a significant impact on the goal of social economic development—typically the impact of policies to reduce emissions in AFOLU on food security. The current efforts in reducing poverty, curbing malnutrition and improving incomes are oriented toward increasing the rate of food production per capita in developing countries, while population growth will require increases in income. Therefore, a policy of increasing food production is needed to ensure sustainable development of the country. Accordingly, solutions to reduce emissions from the AFOLU sector if contributing to food production will contribute positively to this work. In contrast, there will be a number of solutions that may reduce food productivity, at least at the local scale.

In the energy sector, energy scarcity has prompted developed countries to seek biofuel sources, displacing food production in agriculture. This has caused serious food shortages. These factors push up food prices, making the supply drop, and poor countries suffer the most severe consequences.

5.3.4 The Impact on Employment

In the case of policy applied to goods prices based on the corresponding GHG emissions created during their production and consumption, these policies may affect the total supply as well as the distribution of employment between sectors of the economy. For example, the commercialization of emissions to manage emissions from the energy sector of the USA has only generated a small amount of change in total employment over the long term, but the changes created by this policy have partly impacted on employees. Specifically, rising energy prices reduce the real wages of workers. Meanwhile, some people may choose to work fewer hours, or even stop working to switch to operating in other areas.

Although there is no major impact in the long term, the GHG emission reduction policies can cause a significant shift in the structure of labor between sectors of the economy. For example, the commercialization of emissions from the energy sector in the USA could reduce the number of carbon- and energy-intensive industries in energy production or manufacturing of products consuming energy due to these industries facing the problems of increased production costs and reduced outputs. In particular, the energy industry, such as coal mining, oil and gas, may be most severely affected. In addition, emission reduction policies also affect employment in industries that use high-emission products, such as the transportation and chemical industries. In contrast, the policy will create new jobs in other sectors, particularly the manufacturing of machinery to produce energy without CO_2 emissions, such as producing electricity from wind and solar power. Similarly, employment can be increased in sectors producing goods and services using less energy or less energy consumption products, in which the services sector may have the most significant increase.

5.3.5 The Impact on Energy Security

Security of the energy supply side can be defined as "the availability of energy at all times in many forms to ensure sufficient quantity and at an acceptable price" https://www.iea.org/topics/energysecurity/subtopics/whatisenergysecurity/. This definition refers to the prevention and mitigation of emergencies in the short term as well as limiting the risk of energy security in the long term.

Climate change and energy security have become the two main drivers of energy policy in the future of the country. While energy security has been the focus of energy policy for nearly a century, the concern about climate change has emerged in recent times, but has a significant influence and alters virtually all of the context of energy policy. The key problem that decision makers are faced with is how to simultaneously ensure national energy security, while reducing GHG emissions. There is no guarantee of energy security, often due to the non-availability of energy and because energy prices are not competitive or are too unstable. In fact, these effects are often very difficult to assess; therefore, it is difficult to determine a reasonable policy. In the context of climate change, countries usually have a certain number of activities to ensure energy security. Firstly, countries may seek to minimize the short-term effects due to lack of power supply in case of power interruption or, secondly, may make efforts to improve energy security in the long term. In the first case, the country often resolves to build strategic reserves. For example, in the case of oil, the International Energy Agency (IEA) coordinates the use of emergency oil reserves between member states. The government also seeks to establish contingency plans to limit consumption, thereby minimizing the impact due to the lack of energy. In the second case, the policies tend to focus on determination of the root causes of loss of energy security in the context of climate change, such as interruption of power systems related to catastrophes or extreme weather conditions; balancing supply and demand in the market for short-term power; monitoring the effectiveness of management and regulation; and focusing on fossil fuel sources by minimizing the possibility of the risk of depending on the supply of energy in the traditional market and reducing use of fossil fuels or diversifying the type of power supply.

Each system of energy security policy allows identification of potential overlaps with policies and measures to reduce GHG emissions from the energy sector. For example, policies to address resources can significantly affect GHG emissions and vice versa, because they tend to impact the choice of fuel and related technologies. In contrast, policies to overcome regulatory failures can only have a secondary impact on emission reduction policies. Thus GHG emission reduction policies can cause a great impact on plans and strategies to ensure energy security in the country.

5.3.6 The Impact on the Environment

The GHG emission reduction policies may also impact the environment. For example, in the energy sector, hydropower development can affect the environment and ecosystems at the construction site. In the AFOLU sector, emission reduction policies often affect land availability and competition, while land developers may have different perspectives on the importance of ecosystem services. Policies to increase food production may reduce environmental services. Policies to reduce GHG emissions from the agricultural sector often have positive impacts as can be seen in countries that have suffered from declines in water quality and ecology, and sedimentation. These losses can be reduced by implementing conservation tillage measures that will provide benefits in terms of land recovery, or will limit soil erosion. Other positive externalities of GHG reduction policies are changes in farming practices to cause an increase in organic matter in soil, improve the water holding capacity of the soil and reduce the need for irrigation; increased organic matter in the soil can improve soil fertility, which reduces the need to use inorganic fertilizers; conversion from farmland to grassland or forest land can improve the habitat of wildlife and biodiversity protection; restriction of fertilizer use can reduce the nutrient content of the overflow from agricultural land, thereby improving water quality and reducing the shortage of oxygen in rivers, streams, lakes and aquifers. These changes will improve the characteristics of the water used for non-agricultural activities in the area.

However, besides the positive impacts, the greenhouse gas emission reduction policies in the AFOLU sector can also create externalities costs that are not small, as in some cases, reducing the intensity of arable land use requires the use of more pesticides to control weeds, fungi and insects. In addition it requires additional energy for synthesis, production and application. These activities also have negative impacts on the ecosystem, flow and water quality.

5.3.7 *Reducing Costs and Losses from Climate Change Impacts*

The adoption of policies to reduce GHG emissions help to avoid the risks of climate change. Such risks include reduced potential crop yields in most tropical and subtropical regions due to the increase in temperature; reduced and changed crop yields in most regions at mid-latitudes due to the increase in average annual temperature; reduced water supplies in areas of water scarcity, especially in the subtropics; increases in the number of people exposed to vector-borne diseases (such as malaria) and water-borne diseases (such as cholera) and mortality due to heat stress (heat stress, mortality); increases in the risk of widespread flooding of many residential areas due to increased rainfall and sea level rise; and increased energy demand for cooling due to higher temperatures in the summer. The implementation of GHG reduction policies in these areas will help to avoid the costs or potential losses that are caused by climate change impacts.

5.3.8 *The Social Impact*

Social costs to operate and monitor the climate change and reduction of GHG emission programs are often not small, and include labor costs, raw materials, project implementation costs, the cost of raising awareness and compliance with emission standards, the energy accounting program, reducing emission labeling, etc. In case the costs are not included in these specific GHG emission reduction measures, they should be regarded as a form of external costs. Normally, the costs of GHG emission reduction activities are often much higher if they are calculated fully.

As the above analysis shows, to reduce GHG emissions from different sectors, in addition to calculating the direct costs for reducing emissions, there is a need to assess carefully the effects of the policy of emission reduction, especially the negative externalities; these are the indirect costs to be paid in reducing GHG emissions. The choices of priorities, plans and measures to reduce GHG emissions in accordance with the actual conditions of each country require consideration and full evaluation of all of the externalities.

5.4 Selection of Priority Areas and Measures to Reduce Emissions of Greenhouse Gases

5.4.1 Selection of Priority Areas

Identifying priority areas for policy implementation should be based on the criteria that the externalities have been taken into account. The implementation costs criterion represents the economic efficiency of emission reduction countermeasures, usually expressed as the monetary value per unit of CO_2 avoided when implementing these measures. The assumption is that the lower the cost is, the more attractive the options are. Conversely, the choices for emission reduction measures have higher costs that will not be a priority in the early stages and can be implemented later.

The priority policies should have the ability to meet the emission reduction targets of the country. This indicator reflects the level of impact and the ability to contribute to GHG reduction targets of selected sector emission reduction measures. It is usually based on the percentage of CO_2 emissions from the sector compared with the total amount of the CO_2 cut off. The sectors that have higher potential emissions will be placed at greater priority.

The applicability of the policies is an important criterion to prioritize the measures. This reflects the necessity for a change in legislation or institutional systems to enable the successfully implementation of emission reduction measures. Typically, measures that require little change or effects on other policies when being implemented are often easier to implement and therefore will prevail.

There are several factors affecting the applicability and implementation of GHG emission reduction measures. If there are many similar implementing measures that have been done before, it will be better with this experience. Reducing the number of decision makers who have a key role in the implementation of measures to reduce emissions in priority areas would help to implement the measures quickly and effectively. The complexity of the preparatory activities will help to shorten the timeframe for implementation. The level of diversity of the groups that the reduction measures will be directed at should be as little as possible. As usual, a greater number of targeted groups will require more work on related policies.

The reduction measures should have the ability to combine with activities to improve the quality of life. This reflects the extent to which the measures will supplement and support policies and other measures aimed at improving the quality of life of people, such as poverty reduction and energy security. To a certain extent, this reflects the priority that reducing emissions will not only limit the amount of CO_2 emissions but also contribute to reducing the burden of negative effects on people's health or the ecosystem.

Preferable measures will help to create job opportunities. This is the social impact of the measures. It is mainly based on indicators of jobs directly created to implement measures in priority areas to reduce emissions.

Another important factor to be considered in determining the priority areas for reducing GHG emissions is that they need to be in compliance with the priorities of the country's development policies. Determining priority areas for GHG emission reduction therefore is weighted in accordance with the development goals of the country. The priorities for development will be essential in determining areas for reducing GHG emissions.

5.4.2 *Identifying Technical Solutions—Technology Priorities in Reducing Greenhouse Gas Emissions*

After defining the field of emissions and identifying technical solutions, specific technologies to reduce emissions from different sectors also have very important implications for the GHG emission reduction strategy of the country. Usually, technical solutions and technologies are applied to the sector, and sub-sector emission reduction priorities are classified according to their applicability in the short term and long term, or on a large or small scale. The classification results allow comparison with other solutions and building system solutions applied over time. Accordingly, the solution can be applied to many sectors/sub-sectors, though not with the highest priority in all sectors/sub-sectors identified.

5.4.3 *Lessons for Vietnam*

The identification of priority areas to reduce emissions requires specific research, based on national conditions and the development goals. Accordingly, there is a need to develop criteria to identify areas in which to consider the full range of aspects of the potential to reduce emissions, the ability to deploy, the cost and other economic, social and environmental effects.

For reduction of GHG emissions from the energy sector and the promotion and development of new energy sources, renewable energy often requires great support financially from the government to businesses, including investment costs, installation costs, operating costs and R&D activities. Therefore, for developing countries like Vietnam, the budget must also cater for multiple items for other urgent development, and focusing on investment in the development of renewable energy sources will encounter many difficulties, so there is a need for a strategy and roadmap for this. As an agricultural country, Vietnam has the potential for development of bioenergy from agricultural by-products to partially replace fossil fuels to reduce GHG emissions without too much investment cost.

With the economy still relying heavily on agriculture in the next decade, not only for rice but also for other industrial crops (coffee, tobacco, rubber, pepper, and cashews), Vietnam needs to focus on building sustainable agriculture and application of advanced agricultural technologies to maintain and develop the quality and quantity of agricultural production in the context of climate change. For application of this new technology, besides support from the Government of Vietnam, support from developed countries should be enlisted in parallel with relief efforts. This will ensure appropriate orientation and mitigation of GHG emissions associated with climate change adaptation under the most favorable orientation for Vietnam.

In the forestry sector, the strengthening of forestry on vacant land in the tropics is known as an effective measure to reduce CO_2, the main GHG in the atmosphere. In addition to reforestation efforts and reforestation on barren hills, recently the international community has also shown interest in sustainable management of the available forest resources. REDD+ initiatives have been proposed and have received the attention of many nations. This initiative stems from the fact that deforestation and forest degradation contribute a large proportion (15–20 %) of the total amount of GHG emissions due to human activities, and the cause is global in scope.

The formulation and implementation of policies to reduce GHG emissions need to be considered in a comprehensive manner, which requires close coordination between ministries and departments.

The GHG emission reduction strategies should be implemented in a flexible manner, combining policy commands with incentives and economic and technical support to encourage the cooperation of the relevant parties.

Efforts should be made to encourage participation and promote the role of stakeholders in the GHG emission reduction activities. The involvement of the community and stakeholders not only helps to ensure cooperation with and support of government policies but also can help to sustain the motivation of the Government in working toward GHG emission reduction targets.

Leverage, promoting investment, is an important factor in GHG mitigation and response to climate change, especially in developing countries with limited funds. The collaboration and support between Norway and Brazil in the successful campaign to reduce emissions from deforestation in Brazil have demonstrated the role and importance of external sources of support for activities to reduce emissions.

Application of techniques and technology is one of the essential elements to ensure effective policies and activities to reduce emissions, and in return, such mechanisms and policies are promoting the implementation and application of technical solutions and technologies to reduce GHG emission practices.

References

Decision 1393/QĐ-TTg dated 25/09/2012. "Chiến lược Tăng trưởng xanh quốc gia" (2012, in Vietnamese). Available at: http://vanban.chinhphu.vn/portal/page/portal/chinhphu/hethongvanban?_page=1&class_id=2&document_id=163886&mode=detail

General Statistic Office (2014) Yearly Statistics Publication

JICA Inventory Capacity Building Project (2014) Interim report, June 2014. Ministry of Natural Resources and Environment of Vietnam

Low-Emission Development Strategies (LEDS), Technical, Institutional and Policy Lessons. Clapp et al. OECD (2010); Available at http://www.oecd-ilibrary.org/environment/low-emission-development-strategies-leds_5k451mzrnt37-en?crawler=true

Ministry of Industry and Trade (2013) Situation and development direction of environmental technology of Vietnam up to 2020, vision 2030. In: Conference of environmental technology development, 2013, Ministry of Trade and Industry of Vietnam (in Vietnamese)

The Second Communication Report, Ministry of Natural Resources and Environment (MONRE), 2010

UN Data. http://data.un.org/Data.aspx?d=MDG&f=seriesRowID:751

Part II
Bridging the Gap Between Modeling and Real Policy Development

Chapter 6
Designing a National Policy Framework for NAMAs

Lessons Learnt from Thailand

Bundit Limmeechokchai

Abstract This section presents lessons learnt from Thailand in climate policy design. Thailand has filled the gap between modelling analyses and climate policy development in its Nationally Appropriate Mitigation Action (NAMA). Thailand's mitigation pledge under NAMA framework was successfully designed and communicated to UNFCCC in COP20. The integrated assessment modelling analysis plays an important role in the development of Thailand NAMA. Consensus building was derived from several discussions among stakeholders of NAMA implementation. Criteria for selection of greenhouse gas countermeasures were based on cost optimization by using a module of the Asia-Pacific Integrated Model called 'AIM/Enduse', abatement costs, co-benefits and feasibility of implementation. In addition, economic feasibility of countermeasures in NAMA actions was also assessed. Then, NAMA implementation has been prepared based on assumptions concerning limitations of resources, capital requirement, timing and appropriateness for Thailand.

Since 2012 Thailand's mitigation pledge to UNFCCC has been prepared on the basis of domestic appropriate measures. Co-benefits of NAMAs are also assessed, and they reveal positive aspects of GHG mitigation under NAMA framework. Results found that Thailand has high potential of GHG emission reduction by both domestically supported NAMAs and internationally supported NAMAs about 23–73 million tonnes CO_2 per year in 2020 or approximately accounted for 7–20 % in 2020 of the total GHG emissions. The NAMA actions include measures in (1) renewable electricity, (2) energy efficiency, (3) biofuels in transportation and (4) environmental sustainable transport system. These GHG countermeasures are in line with the national policy and plans of ministries of energy and transport in order to avoid the conflict between climate policy and policies of the related ministries. Results of cost optimization, co-benefits, economics and appropriateness are also necessary for communication among policymakers, administrators, academic researchers and the public on consensus building.

B. Limmeechokchai (✉)
Sirindhorn International Institute of Technology, Thammasat University, Bangkok, Thailand
e-mail: bundit.lim@gmail.com

S. Nishioka (ed.), *Enabling Asia to Stabilise the Climate*,
DOI 10.1007/978-981-287-826-7_6

Finally, to ensure the quantified GHG reduction in 2020 and the transparency of Thailand's NAMA implementation, the measurement reporting and verification (MRV) process is required. The MRV process of these NAMAs needs cooperation among related ministries. These lessons learnt from Thailand, when modified as needed, can be a 'good practice' of climate policy design.

Keywords Thailand NAMA • Integrated assessment modelling • Renewable energy • Energy efficiency • Co-benefits of GHG Mitigation • AIM/Enduse

Key Message to Policymakers

- Thailand successfully developed its Nationally Appropriate Mitigation Action (NAMA).
- Integrated assessment modelling helps in climate policy development.
- Consensus building is necessary among stakeholders' concerns.

Lessons learnt from Thailand can be a 'good practice' of climate policy design.

6.1 Introduction

Within the climate change framework, there is a gap between modelling analyses and policy development, and whether national climate policy incorporates such modelling analyses depends on several factors. This section introduces a good example of a means to fill in this gap. Through discussion with the climate change focal point and related agencies, Thailand succeeded in reflecting the modelling analysis in actual policy development. Thailand's scenario studies on Nationally Appropriate Mitigation Action (NAMA) and NAMA roadmap development have been highly successful. This approach can be adopted by other regions as a 'good practice' of climate policy design and modified as needed according to local conditions.

6.2 NAMA and CO_2 Mitigation Strategy

Climate change and greenhouse gas (GHG) mitigation are two key global issues, with a growing list of countries adding them to agendas within United Nation Framework Convention on Climate Change (UNFCCC) discussions. Further, the Conference of Parties (COP) has decided on appropriate implementations and GHG mitigation targets for developing countries. The 'Nationally Appropriate Mitigation

Action (NAMA)' concept was first introduced in the 'Bali Action Plan' in COP13 in 2008 and, for developing countries, involves submission of GHG mitigation targets at the request of COP. In other words, the targets are only GHG mitigation 'pledges' (Decision 1/CP.13 'Bali Action Plan', Decision 1/CP.16 'Cancun Agreements', Decision 2/CP.17 in Durban and Draft, Decision -/CP.18 'Doha Climate Gateway'). Moreover, developing countries are welcome to propose their actions and targets for GHG mitigation under the voluntary basis of the UNFCCC. As of October 2012, 54 countries had proposed mitigation pledges that included NAMA implementation. Thailand communicated its NAMA pledge to UNFCCC in COP20 in Lima (Dec. 2014).

In the convention-track decision, developing countries agree to take on NAMAs, supported by technology and finance, based on their goal as being 'aimed at achieving a deviation in emissions relative to 'business-as-usual' emissions in 2020'. Developed countries are urged to raise ambition levels of their targets 'to a level consistent with' the latest recommendations of the Intergovernmental Panel on Climate Change (IPCC). Developed countries have been requested to prepare 'low-carbon development strategies or plans', and so have developing countries.

The term 'NAMAs' refers to any national climate policy that leads to reduction in greenhouse gas (GHG) emissions in developing countries. The mitigation pledges communicated to UNFCCC by signatory countries of NAMA agreements can be classified into four main groups: (1) NAMA concept, (2) NAMA plan, (3) NAMA implementation and (4) NAMA submitted to the UNFCCC's NAMA registry. Thailand's NAMAs are in line with national sustainable development plans and are aimed at achieving GHG emissions reduction relative to 'business as usual' emissions in 2020, resulting in GHG mitigation. NAMAs have impacts that can be measured, reported and verified (MRV) and comprise two types in Thailand: (1) domestically supported NAMAs and (2) internationally supported NAMAs. Both types require MRV processes to verify actual emission reductions and to provide transparency of the processes. In 2014, Thailand constructed a national strategy 'Roadmap to Thailand NAMAs 2020' with clear targets to set up benchmarks and orient emission reductions activities. In December 2014, the country announced its NAMA pledge with a GHG reduction target in the range of 7–20 % in COP20 in Lima. The pledge communicated to UNFCCC was approved by the cabinet, the national climate change committee and NAMA subcommittee and stakeholder consultations. As this pledge has no legal binding for the country, the final NAMA pledge process only needed approval at the cabinet level (see Fig. 6.1), and if it were legally binding, it would have required parliamentary approval, in accordance with the constitution. In such case, the Department of Treaties and Legal Affairs of the Ministry of Foreign Affairs would be involved. If the post-2020 mitigation pledge involves legal binding, the domestic processes leading up to communication to UNFCCC will require more time.

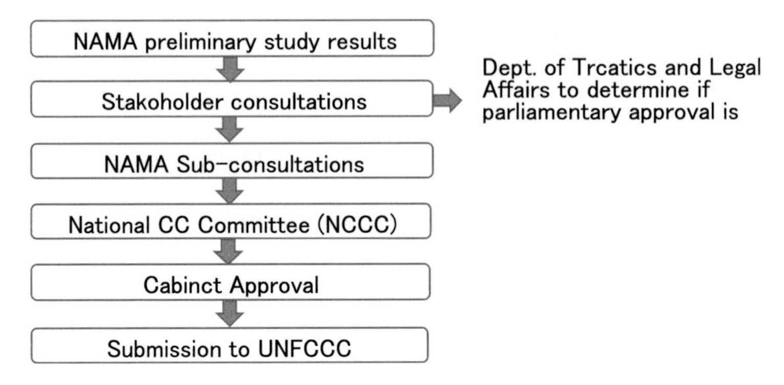

Fig. 6.1 National NAMA approval process of Thailand

6.3 Implementation of a Climate Change Mechanism in Thailand

In 2012, the first study by Thailand Greenhouse Gas Management Organisation (TGO) showed that Thailand has a high potential for GHG emissions reduction, via both domestically and internationally supported NAMAs, to the tune of about 23–73 million tonnes CO_2 per year by 2020 or approximately 7–20 % of the total GHG emissions for 2020. The calculated abatement costs of NAMAs vary in the range 0–1000 USD/t-CO_2. However, these CO_2 reduction actions are voluntarily for Thailand. In order to convey its intent as regards being a main supporter for climate change reduction and GHG mitigation in SE Asia, Thailand has to be ready for the coming strategies in the proposed NAMAs. The TGO working group and the Ministry of Energy found that the national sectoral approach actions are composed of:

1. Alternative Energy Development Plan (AEDP) 2012–2021 (updated in May 2012)
2. Energy efficiency measures in industrial and building sectors, according to the act on energy conservation passed by the Ministry of Energy
3. Biofuel promotion in the transport sector under AEDP
4. The mass transit system, termed 'environmental sustainable transport system', by the Office of Transport and Traffic Policy and Planning (OTP)

Thailand's mitigation pledge to UNFCCC has been prepared on the basis of these measures since 2012. In addition, co-benefits of NAMAs are also assessed, and they reveal positive aspects of GHG mitigation under the NAMA framework. The MRV process of these NAMAs requires cooperation among the related ministries.

6.3.1 Success of Clean Development Mechanism (CDM) Projects in Thailand

As of January 2012, TGO reported that the number of CDM projects issuing CERs had increased to ten, with a total CO_2 reduction of 1.05 Mt-CO_2. However, as developing CDM projects up to the issuance of CERs takes much time, it is recommended that Thailand's NAMA process be made more flexible and simplified to reduce the overall processing time and that other developing CDM projects be switched to register in the NAMA category. It is anticipated that Thailand will fast-track projects under NAMAs. As of December 2014, 222 Thai CDM projects had received the letter of approval (LoA), accounting for a cumulative GHG reduction of 12.72 Mt-CO_2. Of these, however, only 151 projects had been registered by the CDM Executive Board (CDM-EB) (with a cumulative GHG reduction of 7.25 Mt-CO_2), and only 43 projects had received issuance of certified emission reductions (CERs) (equal to 6.92 Mt-CO_2).

6.4 Overview of Energy, Environment and Socio-Economic Factors

6.4.1 Thailand's Energy Sector

Thailand is the second largest economy in the Association of Southeast Asian Nations (ASEAN) and is categorised as a middle-income developing country. Since recovering from the Asian financial crisis in 1997, its economy has shown significant growth, rising to 224 billion USD in 2012 and growing on average at 4.07 % annually. Medium-term economic projections point to Thailand maintaining this growth rate (about 4 % annually), and one report ('Thailand Power Development Plan 2010', published by Ministry of Energy, Thailand) quotes a figure of 4.27 % for 2012–2030.

The population of Thailand expanded 2.4 times during 1960–2012 and reached 66.8 million in 2012. Two decades ago the average annual population growth rate dropped to 0.74 % and then to 0.46 % in the most recent decade. The urban population was reported as 34.4 % in 2012. Accompanying the expected gradual economic growth, infrastructure development, increased access to modern energy and high-income generation will lead to increased urbanisation in future; thus, the urban population is forecasted to rise to 60 % by 2050.

6.4.2 *Primary Energy Supply and Final Energy Consumption*

Behind Indonesia, Thailand is the second largest energy consumer in ASEAN and consumed over 70.5 Mtoe in 2011, amounting to just over 20 % of regional demand. Of this, 70 % was due to transport and industrial sectors, with the transport sector being the largest energy consumer in 2011 (25,466 ktoe), closely followed by the industrial sector (24,966 ktoe). The residential sector is the third (16,551 ktoe). Over the last two decades, the highest average annual growth rate in energy demand was recorded in the industrial sector at 6.78 %, followed by transport and residential sectors at 5.16 % and 3.1 %, respectively.

In terms of fuel mix, as shown in Fig. 6.2, petroleum products dominate the total final energy consumption (TFEC) at 47 %, followed by electricity at 18 %. The dependency on fossil fuels is significant, as petroleum products, coal and its products and natural gas amounted to 63 % of TFEC in 2011.

As regards total primary energy supply for the world, it doubled from 1973 to 2011. Figure 6.3 gives a comparison of fuel mix in 1973 and 2011. According to the figures, fossil fuels have maintained their dominance since the 1970s. Except for the increase in nuclear share, from 0.9 % in 1973 to 5.1 % in 2011, none of the non-fossil fuel shares exhibit significant increases, though some of them have higher incremental factors, such as solar PV and wind power. In addition, the contribution of oil to total primary energy supply (TPES) dropped from 46 to 31.5 %, whilst natural gas and coal increased to 21.3 % and 28.8 % in 2011, from 16 % and 24.6 % in 1973, respectively.

In 2011, TPES in Thailand was recorded at 127.9 Mtoe, whilst still only 0.97 % of the world and 8 % of Asia excluding China attained this figure due to an average annual growth rate of 6.74 % since 1990. As with the global statistics, oil has the major share in the fuel mix of TPES at just above 36 %, followed by natural gas at 32 % and biomass at 16.2 % in 2011. The natural gas share has grown considerably due to its increased use in power generation. On the other hand, growth in energy from coal has been restricted due to its environmental concerns in Thailand and had a share of about 12.3 % of TPES in 2011. However, the percentage for fossil fuels in TPES stood at about 85 % in 2011.

6.4.3 *Thailand's GHG Emissions*

In 2000 Thailand emitted about 229 Mt-CO_2eq, most of which is due to the power sector. It is followed by transport and industry sectors, each accounting for 44.4 and 30.3 Mt-CO_2, respectively. Figure 6.4 shows sectoral shares of CO_2 emissions in 2000 and in particular shows that the energy sector is the biggest CO_2 emitter, accounted for 69.57 % of the total. The share of CO_2 emissions in the agricultural

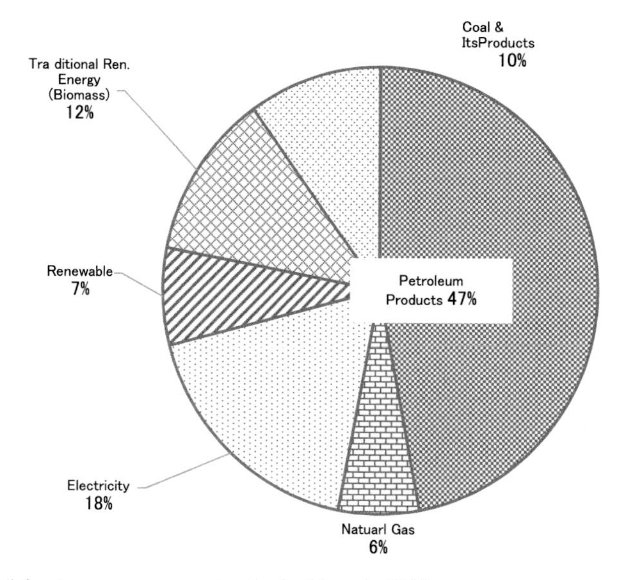

Fig. 6.2 Total final energy consumption by fuel type in 2011

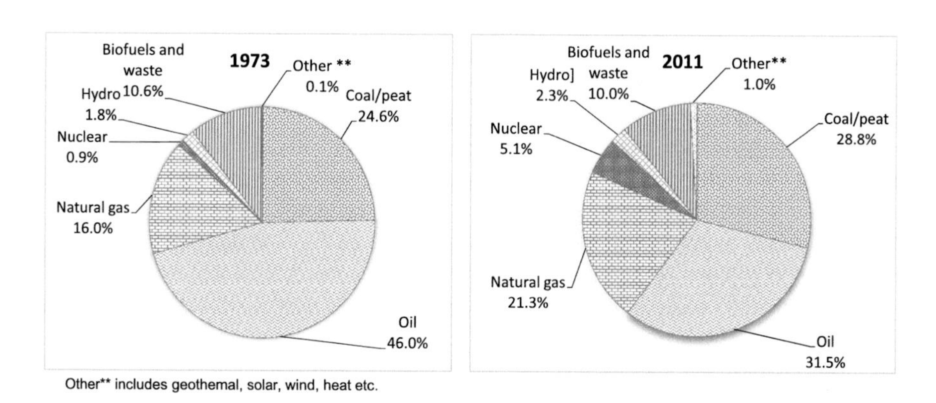

Fig. 6.3 Global fuel share of TPES for 1973, 2011

sector is about 22.64 %. The forestry sector and carbon sequestration result in CO_2 absorption of 3.44 %.

The trend in Thailand's GHG emissions shows an increase—in 1992 total CO_2 emissions stood at about 100,033 kt-CO_2, which then increased to 194,853 in 2009, and accounted for an average increase rate of 2.33 % per annum. The biggest CO_2-emitting sector is the power sector, which was responsible for 41,838 kt-CO_2 in 1992, and 81,797 in 2009, and accounted for an average increase rate of 5.14 % per annum. This is followed by the industrial, transport and building sectors, which accounted for average annual increase rates of 1.66 %, 2.00 % and 2.98 %, respectively.

In 2011, CO_2 emissions of 206.4 Mt-CO_2 were recorded, a contribution of only 0.66 % to the global figure. However, CO_2 increased by 202.4 % over the last two

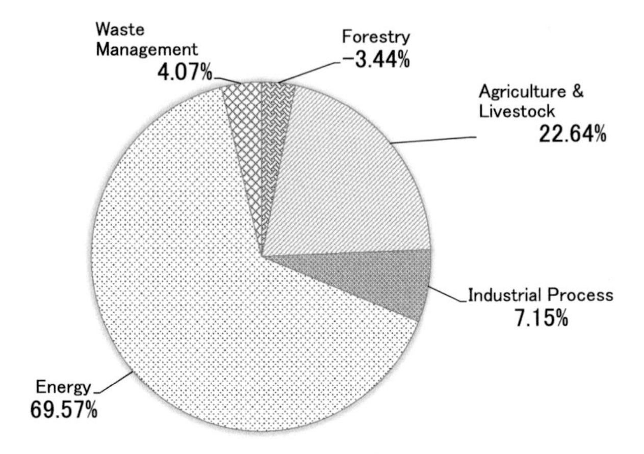

Fig. 6.4 Breakdown of CO_2 emissions by sector in 2000 (Source: Thailand's Second National Communication, ONEP 2011)

decades. Of the total, 41.8 % came from the power sector in 2011, and second was transport at 28.52 % (58.87 Mt-CO_2). However, although the emission quantity is still comparatively small, emissions from residential and commercial sectors have the highest average annual growth rate of 5.35 %, followed by the industry and the power sectors of 4.47 % and 3.89 %, respectively. A fuel-wise comparison of carbon emissions shows that by far oil is the largest emitter and was responsible for about half of CO_2 emissions for 2011. Due to its high use in power generation, natural gas has also incurred considerable emissions (about one third of the total) followed by coal (about one fifth).

In the energy sector, the energy conversion processes in electricity power plants are the chief contributor to CO_2 emissions, which are followed by combustion processes in industry and transportation. During 2002–2010, the corresponding CO_2 emissions increased by 30.03 %, 18.78 % and 12.47 % in the power, industry and transport sectors, respectively. Therefore, it is recommended that related policy measures in GHG mitigation in the energy sector be focused on energy conversion processes and fossil fuel combustion processes in industry and transportation sectors. Thus, for Thailand's NAMA, the first study focused on power generation, industry and waste to energy activities.

6.4.4 Other Air Pollutant Emissions

NO_x emissions in 2011 stood at 971 kt-CO_2 and during the period 1994–2011 increased by about 83 %. The transport sector is the major NO_x emitter in Thailand and in 2011 was responsible for 307 kt-CO_2, an approx. 31.6 % share of the total. However, power, manufacturing and others (agricultural, construction and mining) also emitted considerable amounts.

With regard to SO_x emissions, the figures dropped considerably during the period 1994–2011, from 1326 kt in 1994 to 552 kt in 2011. Since the power sector is responsible for the highest SO_x emissions, desulphurisation retrofits in power plants have decreased SO_x emissions drastically. However, in 2011 the power sector emitted the highest amount at about 55 %. More than 96 % of SO_x emissions in 2011 were produced from the power and manufacturing sectors.

6.5 Relationship Between Thailand's Energy Policy and Climate Change

The energy policy of the Ministry of Energy of Thailand has three main strategy aims: energy security, promotion of alternative energy and increased energy efficiency in the enduse sectors, as follows:

1. The energy security strategy aims at development and promotion of endogenous energy resources to satisfy domestic consumption. It includes cooperation with nearby countries in the development of energy sources and utilisation of renewable energy resources such as small hydroelectricity generation, development of nuclear power and utilisation of clean coal technologies.
2. The alternative energy strategy was announced by the Thai government as a national agenda in 2008. Promotion and utilisation of alternative energy supplies include biofuels, biomass, biodiesel and waste to energy. In addition, renewable electricity generation from solar, wind and hydro has been targeted in the 15-year Alternative Energy Development Plan (AEDP) of the Department of Alternative Energy Development and Efficiency (DEDE), Ministry of Energy. The first 15-year AEDP plan made in 2011 is divided into three periods: 2008–2011, 2012–2016 and 2017–2022. This plan will result in cumulative energy savings of 19,799 ktoe by 2022 and will account for 20 % of total final consumption in 2022.
3. The increased energy efficiency strategy of DEDE aims at targeting energy savings in the residential, industrial, commercial and transport sectors. It includes promoting awareness and understanding of energy conservation, providing financial incentives to the private sector in retrofits of energy equipment for energy savings, peak load cutting in the commercial sector and R&D on minimum energy performance standards, building energy codes and mass transit systems. Efficient use of energy results in less investment in the energy supply (DEDE, 2009).

6.5.1 Revised Alternative Energy Development Plan: AEDP 2012–2021

As mentioned above, the first AEDP is divided into three subperiods and aims to promote renewable electricity generation such as biomass power, biogas power,

Table 6.1 Potential of renewable power generation in the AEDP plan

	AEDP plan (MW)[a]		Revised in 2012[b]
Renewable energy	2008–2011	2017–2022	2021 (MW)
Solar	55	500	3000
Wind	115	800	1800
Wastes to energy	78	160	400
Biomass	2600	3700	4800
Small hydro	165	324	324
Biogas	60	120	3600
Others	–	–	3
Total	3073	5604	13,927

Source: [a]DEDE (2010)
[b]DEDE (2012)

small hydropower, solar power, wind power and waste to power. In January 2012, DEDE increased the target of 2012 AEDP to 25 % by 2021. To achieve this target in the revised AEDP plan, Ministry of Energy has provided financial support and mechanisms to promote renewable electricity generation in the form of 'adders' on the top of buy-back rates. Adders of different renewable energy sources will be given by different rates. The concept of these incentives is to make the renewable energy investment yield sufficient profits within given lifetimes under specified economic criteria. The target of renewable electricity generation in its first AEDP plan was 5604 MW, in which biomass power will share about 3700 MW and will account for 66 % of the total capacity by 2022 (see Table 6.1).

In 2012, the National Energy Policy Committee (which includes the prime minister) announced the revised AEDP plan. Total renewable power generation was to be 13,927 MW in 2021 with expected annual energy generation of 63,035 GWh in 2021. The largest share of power capacity will be represented by 4800-MW from biomass electricity generation (see Table 6.1).

6.5.2 Thailand's 20-Year Energy Efficiency Development Plan

In 2011, the Ministry of Energy announced the 20-Year Energy Efficiency Development Plan (EEDP). This plan aims at cutting off overall energy intensity and total final energy consumption by 25 % and 20 %, respectively, in 2030. The main sectors targeted are transport and industry (EPPO 2011). EEDP will provide both mandatory actions and voluntary support to promote energy efficiency. The mandatory actions will comply with the 1992 Energy Conservation Act (ENCON Act) and the revised 2007 ENCON Act and minimum energy performance standards including energy labelling. The main form of support to promote energy efficiency is financial, such as incentives for measures proven to enable energy savings, incentives to promote the use and production of energy devices that comply with minimum energy performance standards and incentives to promote high energy efficiency vehicles.

6.5.3 Thailand Power Development Plan (PDP) 2010–2030

The Thailand Power Development Plan is the government-approved development pathway of the power sector for the period 2010–2030. Two objectives have been accomplished by developing this plan. First, the future peak power demand and the total energy demand have been forecasted. Second, a roadmap for power generation expansion to meet forecasted demand has been developed. The PDP 2010 plan has been revised several times for a variety of reasons and changes in national energy circumstances. The first revision was made since the peak demand in 2010 was significantly higher than as forecasted and capacity addition of independent power producers (IPPs) was lower than planned due to delays in plant construction. Then, due to lowered public acceptance for nuclear power as a result of Japan's Fukushima incident, a second revision was prepared. The third and latest revisions were prepared based on three key issues: (1) to adopt forecasted power demand results, (2) to include the guidelines given in the revised Alternative Energy Development Plan (AEDP 2012–2021) and (3) to place more emphasis on energy security.

In PDP2010 Revision 3, to forecast power demand, energy saving programmes and energy efficiency promotions have been considered at a success rate of only 20 %, in accordance with MoEN's (Ministry of Energy) 20-Year Energy Efficiency Development Plan 2011–2030. According to the forecasts, the peak generation requirement in 2030 is 52,256 MW, and net energy generation requirement is 346,767 GWh. These figures have been set as the baseline for preparation of the PDP. In addition, the following features can be highlighted. One of the considerations taken into account was keeping the reserve margin at the level of 20 % higher than the peak demand due to risk in the natural gas supply sources in western Thailand. In addition, diversification of fuel has also been considered to reduce the natural gas dependency. Another consideration was to maintain the share of nuclear power below the 5 % margin. In the plan, a capacity of 2000 MW has been planned for nuclear in 2026. Increasing the share of renewable power generation by 5 % from the level proposed in its PDP Revision 2 has been taken into account. In the PDP 2010 plan, adding 14,580 MW of renewable power capacity has been planned for the system, 9481 MW coming from domestic sources and 5099 MW via purchases from neighbouring countries. Promotion of cogeneration and increasing the power purchased from it were another assumption made, with cogeneration accounting for 6476 MW of capacity addition. In addition, 25,451 MW from combined cycle power plants and 8623 MW thermal power capacity, including 4400 MW of coal, have been added for the period 2012–2030. Altogether, 55,130 MW of new capacity has been added for 2012–2030, whilst 16,839 MW of capacity has been retired. At the end of 2030, net operating capacity stands at 70,686 MW for 2030.

6.5.4 *Environmental Sustainable Transport System*

The 'environmental sustainable transport system' was proposed and developed by the Office of Transport and Traffic Policy and Planning (OTP). The actions, which mainly involve a modal shift, fuel economy improvement and sustainable mass transit system, started in 2012. However, due to the long lead time for construction, this sustainable system is not due to be fully operational until after 2020.

6.6 AIM/Enduse Modelling of Thailand's Energy System

The methodology used to develop Thailand's NAMAs is based on a bottom-up tool, the 'AIM/Enduse' model. The Asia-Pacific Integrated Model (AIM) was developed by the National Institute for Environmental Studies (NIES) of Japan as the first and only integrated assessment model specific to Asia and was used to evaluate policy options on sustainable development particularly in the Asia-Pacific region (Kainuma et al. 2003). AIM/Enduse is a bottom-up optimisation model comprising a detailed technology selection framework within a country's energy-economy-environment system. It can analyse mitigation scenarios by using both the AIM/Enduse model and AIM/Enduse tools. In the model, 'energy technology' refers to a device that provides a useful energy service by consuming energy. Energy service refers to a measurable need within a sector that must be satisfied by supplying an output from a device. It also can be defined in either tangible or abstract terms; thus, 'service demand' refers to the quantified demand created by a service; i.e. service outputs from devices satisfy service demands. The AIM/Enduse leader in Thailand is Prof. Ram M Shrestha, who has developed and analysed climate policies for several Asian countries, as well as Thailand.

In this study, the structure of AIM/Enduse for Thailand's reference energy system was created using socio-economic assumptions obtained from related agencies such as the Office of the National Economic and Social Development Board (NESDB), Electricity Generating Authority of Thailand (EGAT 2010) and DEDE (see Fig. 6.5). Then, the selected CO_2 countermeasures are analysed. The AIM family tools could handle the problems in both general equilibrium and partial equilibrium modelling. Thus, the AIM/Enduse tool is highly suitable for analyses of CO_2 countermeasures in Thailand's NAMAs. Results from the AIM/Enduse modelling for Thailand's energy system show that final energy consumption by economic sectors in the BAU will increase from 71,491 ktoe in 2005 to 113,384 ktoe in 2020 (see Fig. 6.6).

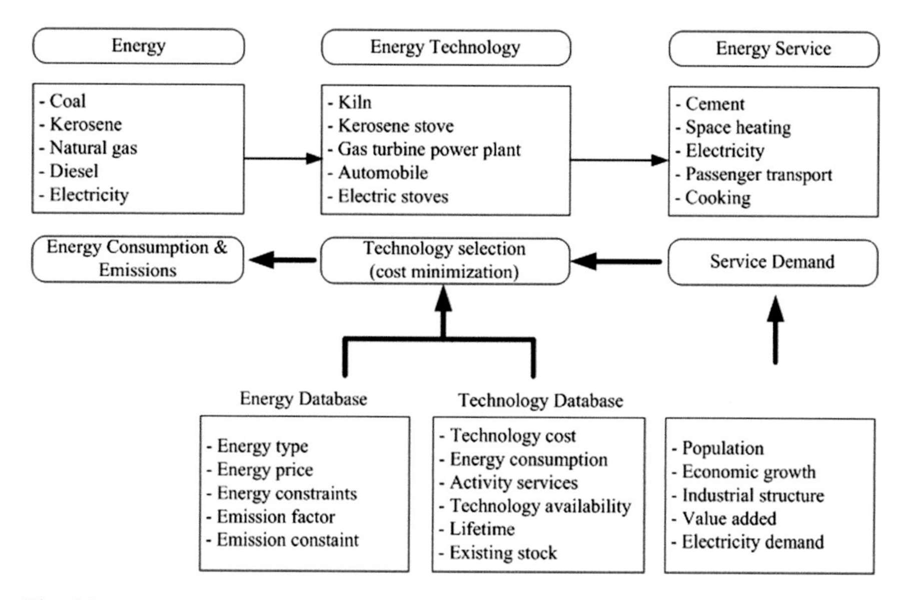

Fig. 6.5 Structure of AIM/Enduse for modelling of Thailand's NAMAs

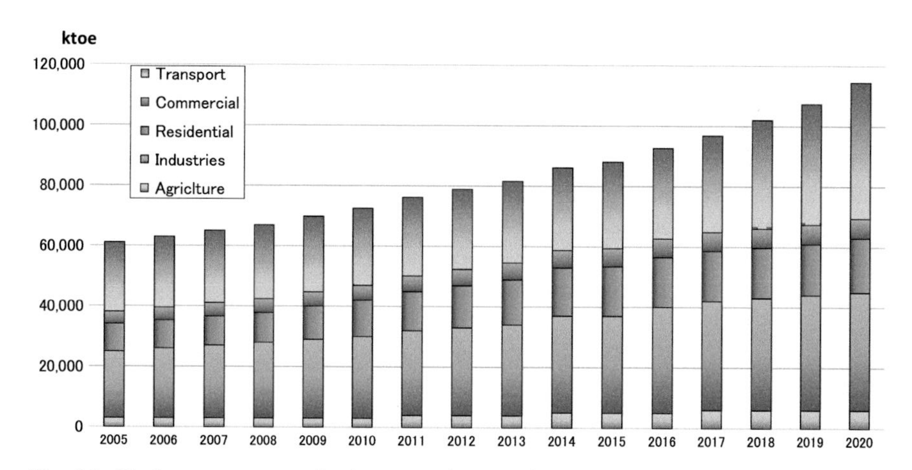

Fig. 6.6 Final energy consumption by economic sector in BAU

The steps of modelling analyses of Thailand's NAMAs are as follows: (1) reviews of national policy measures related to CO_2 countermeasures, (2) data collection and verification, (3) data processing, analyses and modelling by the AIM/Enduse tool, (4) development of CO_2 emissions as baseline in the business-as-usual (BAU) scenario, (5) analyses of CO_2 countermeasures and (6) discussion and conclusion on CO_2 countermeasures under NAMAs (Limmeechokchai et al. 2013).

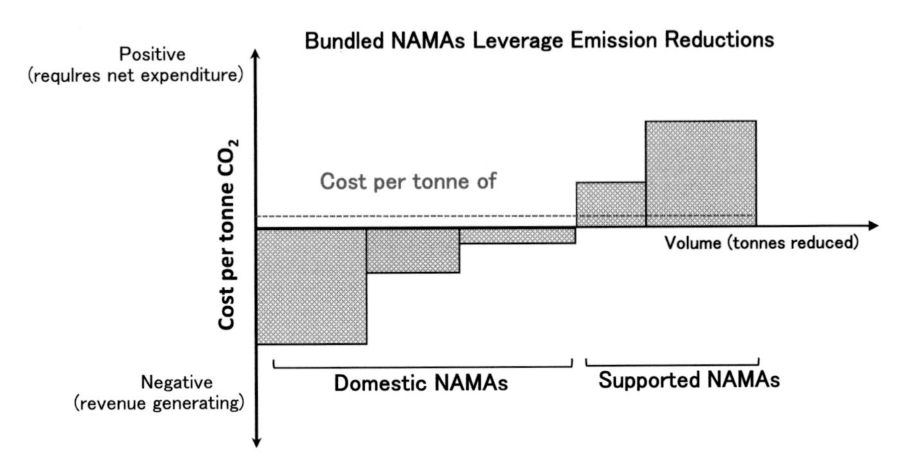

Fig. 6.7 Criteria to selection of Thailand's NAMAs

6.7 Designing a National Policy Framework for Thailand's NAMAs

6.7.1 Criteria/Selection of CO_2 Countermeasures

In the analyses of CO_2 countermeasures (CMs) for Thailand's NAMAs, the abatement costs of selected countermeasures from the national policies and plans, including their economic feasibility, were estimated. The selected countermeasures with appropriate abatement costs were proposed as measures for Thailand's NAMAs (see Fig. 6.7). The proposed countermeasures for Thailand's NAMAs were also assessed for their economic feasibility.

6.7.2 Domestically vs. Internationally Supported NAMAs

The CO_2 countermeasures (CMs) obtained from the AIM/Enduse analyses can be classified into two types:

1. Domestic NAMAs performed voluntarily by the Thai government, sinceCO_2 abatement costs are not excessive and the CO_2 CMs could utilise domestic technologies and know-how
2. Internationally supported NAMAs which have high abatement costs and need support, technology transfer, know-how and capacity building from developed countries

Table 6.2 Internal rates of return (IRR) of CO_2 CMs in renewable power generation

NAMAs	CO_2 countermeasures (CMs)	Incremental abatement costs (\$/t-CO_2)	IRR with adders/FiT (%)	IRR without adders/FiT (%)
Domestic NAMAs	Biogas power	0.02	14.0	8.8
	Small hydropower	0.69	11.7	5.4
	Biomass power	2.67	11.3	4.0
Internationally supported NAMAs	Wind power	51.88	10.8	1.5
	Solar power	102.81	9.0	n.a.

Note: Thailand implemented incentives for renewable electricity generation in its 'adder' scheme in 2008 and 'feed-in Tariff' scheme for solar PV in 2012

6.7.3 *Economic Assessment of Domestic and Internationally Supported NAMAs*

In addition to the incremental abatement costs of CO_2 countermeasures, the internal rate of return (IRR) of each identified CO_2 CM has been analysed. Table 6.2 presents the IRRs of CO_2 CMs for renewable power generation. The CO_2 CMs with high incremental abatement costs tend to have low IRRs, such as the IRRs of wind and solar power without incentives or 'adders' of only 1.5 % and −5.5 %, respectively. When the adders are taken into account, the IRRs of wind and solar power increase to 10.8 % and 9.0 %, respectively. Therefore, both wind and solar power should be considered as internationally supported NAMAs and since need international support to promote the nationwide use of such technologies as CO_2 countermeasures.

On the other hand, renewables powered by biogas, small hydro and biomass have low incremental abatement costs of 0.02, 0.69 and 2.67 US\$/t-CO_2, respectively. In addition to incremental abatement costs, their IRRs without adders are 8.8 %, 5.4 % and 4.0 %, respectively. When 2011 adders are taken into account, their IRRs increase to 14.0 %, 11.7 % and 11.3 %, respectively (see Table 6.2). These IRRs are sufficient and result in financial viability for the renewable power producers in Thailand. Therefore, biogas, small hydro, and biomass power must be classified as domestic NAMAs in Thailand.

For the waste-to-energy sector, from the IRR point of view the MSW of local landfill technology is the best CO_2 countermeasure. However, the identified four MSW technologies show negative IRRs. The IRRs without 2011 adders for MSW-Local landfill, MSW-INC, MSW-BD and MSW-Controlled landfill are −1.3 %, −4.5 %, −6.0 % and −8.0 %, respectively (see Table 6.3). The corresponding incremental abatement costs are 32.85, 140.63, 164.73 and 395.32 US\$/t-CO_2, respectively. When adders were taken into account, their IRRs increase to 11.0 %, 9.0 %, 9.0 % and 10.0 % for MSW-Local landfill, MSW-INC, MSW-BD and MSW-Controlled landfill, respectively. Therefore, all CO_2 countermeasures in

Table 6.3 Internal rates of return (IRR) of CO_2 CMs in the waste-to-energy sector

NAMAs	CO_2 countermeasures (CMs)	Incremental abatement costs (\$/t-CO_2)	IRR with adders (%)	IRR without adders (%)
Internationally supported NAMAs	MSW (local landfill)	32.85	11.0	n.a.
	MSW (incinerator)	140.63	9.0	n.a.
	MSW (digesters)	164.73	9.0	n.a.
	MSW (controlled landfill)	395.32	10.0	n.a.

the waste-to-energy sector will be considered as internationally supported NAMAs since they need financial incentives, technology transfers and capacity building.

For energy efficiency (EE) countermeasures, the payback periods of EE lighting, EE cooling and EE motors in industry were calculated (see Table 6.4). It was found that payback periods of the proposed countermeasures for EE NAMAs in industry are satisfactory. Their short payback periods for business investment are only 3.0–3.5 years. These results are consistent with the stakeholder consultation.

6.8 Framework for Thailand's NAMAs

6.8.1 NAMA Pledge to UNFCCC

Thailand is assessed as having a high potential for GHG mitigation from both domestically supported NAMAs and internationally supported NAMAs, up to 72.99 Mt-CO_2. Of this, 23.33 Mt-CO_2 will be from domestically supported NAMAs and 49.66 Mt-CO_2 internationally supported, as shown in Table 6.5. Figure 6.8 shows the potential CO_2 reduction from Thailand's NAMAs in 2020.

With respect to the potential for GHG mitigation from these measures in 2020, Thailand will be able to provide a draft mitigation pledge as a NAMA concept. Such information includes the base year and the potential of GHG mitigation when compared with the GHG emissions in the target year, 2020.

Thailand's mitigation pledge as a NAMA concept was finalised in 2013. However, more analysis needs to be conducted on the policies in different sectors as well as examples of mitigation pledges submitted by other countries, e.g. China, India, Indonesia, Brazil, Mexico and Chile before drafting Thailand's NAMA policies for the UNFCCC. This step is conceptually important in the development of NAMAs for the implementation phase. It is also beneficial to the country in terms of international financial support for the internationally supported NAMAs.

In conclusion, the total appropriate GHG mitigation by NAMA measures in the energy sector under domestically and internationally supported NAMAs in 2020 will be about 73 Mt-CO_2, accounting for 20 % from the total GHG emissions

Table 6.4 Payback period of energy efficiency (EE) measures in industry

EE in industry	CO_2 countermeasure	Payback period
Domestic NAMAs	EE lighting	3.5 years
	EE cooling	3.0 years
	EE motors	3.0 years

Table 6.5 Proposed CO_2 countermeasures for Thailand's NAMAs in 2020

NAMAs	CO_2 countermeasures	CO_2 reduction in 2020 (kt-CO_2)
Domestically supported NAMAs	RE power	2568
	EE industries	4762
	Building codes	5909
	Transport/ethanol (AEDP 2012)	5069
	Transport/biodiesel first Gen (AEDP 2012)	5022
	Subtotal	23,330 kt-CO_2
Internationally supported NAMAs	RE power (high abatement costs)	13,456
	EE industries (high abatement costs)	9743
	Transport/biodiesel second Gen (AEDP 2012)	14,459
	Environmental sustainable transport (OTP)	12,000
	Subtotal	49,658 kt-CO_2
Total domestic and supported NAMAs		72,988 kt-CO_2
Total emissions in 2005		192,724 kt-CO_2
Total emissions in BAU2020		367,437 kt-CO_2

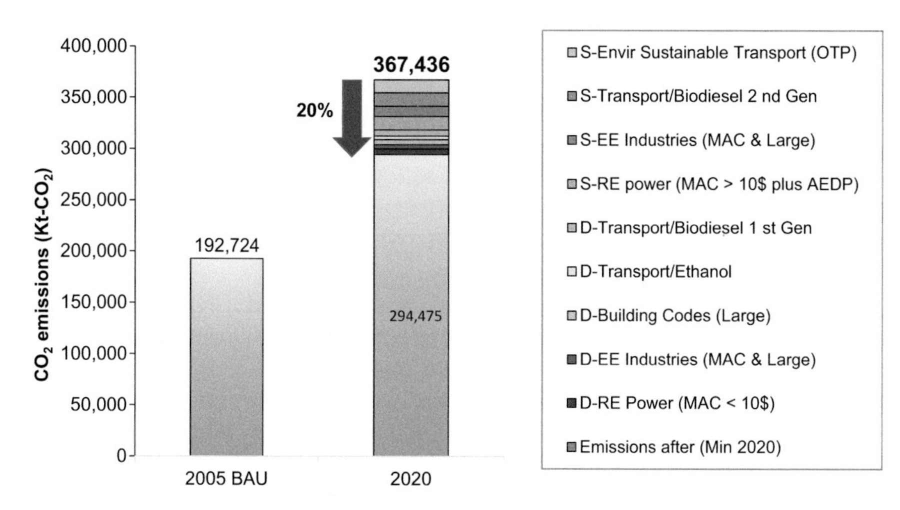

Fig. 6.8 Potential of CO_2 reduction in Thailand NAMA 2020

estimated in the 2020 BAU, where the measures should be explained according to NAMA that:

> Thailand will lower CO_2 emissions by 20 % when compared to the BAU 2020 level.

The potential to reduce the GHG emissions by domestically supported NAMAs is one important measure which could demonstrate the effort Thailand is making as a developing country involved in global GHG mitigation, but, naturally, different countries have different limitations and motivations in developing domestically supported NAMA actions.

A further development in domestically and internationally supported NAMAs is to group NAMAs together, i.e. create 'Bundled NAMAs'. For example, the Power Development Plan (PDP), waste to energy plan, increased energy efficiency in industrial and building sectors plan, promotion of bioenergy in the transport sector plan and increased energy efficiency in the transport sector plan could be combined as a bundle of NAMAs under Thailand's NAMAs, which would demonstrate the importance of GHG emission reductions and funding sources. Thailand has considered carrying out further studies on the environmental impacts in other dimensions, sustainable development and financial availability of the government funds to achieve the 2020 GHG mitigation target.

6.8.2 *Seeking Financial Support*

In Thailand, it is found that there are two sources of domestic funds for the GHG mitigation in energy sector. The first is the 'Energy Conservation Promotion Fund' (ENCON Fund), which was established by the energy conservation act to provide financial support for the implementation of energy security and the development of renewable energy. It is responsible for funding efficiency improvements, renewable and alternative energy development, R&D projects, human resources development, public education, campaigns and environmental projects. The ENCON Fund is currently focused more on projects that follow the Thailand 20-Year Energy Efficiency Development Plan 2011–2030 (EEDP) and Thailand 10-Year Alternative Energy Development Plan 2012–2021 (AEDP) but not the climate change issue.

The second fund is the 'Energy Service Company Fund' (ESCO Fund), which is supported by the Ministry of Energy. Its objective is to encourage private investment in renewable energy and energy efficiency projects. The ESCO Fund is financed by the ENCON Fund to encourage six kinds of investment: (1) equity investment, (2) ESCO venture capital, (3) equipment leasing, (4) credit guarantee facility, (5) technical assistance and (6) assistance for renewable energy projects in selling carbon credits. Due to its benefits, ESCO would appear to be a very worthy mechanism for the domestically supported NAMA plan.

However, as investments for projects tend to be very high, more funds should be provided to help operators. Such projects warrant further study, though procedures for requesting assistance from the ENCON Fund are still very strict and complex due to the limited and discontinuous nature of the fund, which affects motivation and which in turn is why implementation of NAMAs does not go according to plan or may not achieve the targets. Therefore, implementation of NAMAs will have to rely on funding and capacity building from international sources, such as the Green Climate Fund (GCF). It also needs the cooperation of developed countries such as Germany, despite the presence of limitations, complexity and difficulty in accessing the funds. The funding for GHG mitigation actions in Thailand has very high potential of necessity in the near future since domestic funds are not sufficient and there exist barriers for domestic funding.

In addition to RE and EE NAMA actions, Thailand's transport master plan, the 'environmental sustainable transport system', which is a capital intensive plan proposed by the Office of Transport and Traffic Policy Planning (OTP), will also require international support. However, due to the long lead times of the transport system, this system will contribute less in the NAMA 2020 period but will play a key role in GHG reduction potential post-2020.

6.9 Building Consensus Among NAMA Stakeholders in Thailand

During 2012–2013, several NAMA workshops were carried out among stakeholders of GHG emitters in the power, industry and waste to energy and transport sectors to identify the appropriate GHG mitigation measures and potential in Thailand. As regards Thailand's NAMA stakeholder workshops, the steps in organising stakeholder workshops by ONEP (2014) and TGO (2014) during 2012–2013 are as follows:

1. Organising the workshops and work with experts in related GHG mitigation sectors in order to prepare GHG mitigation potential and mitigation plans in the power, industrial, transport and waste to energy sectors, including preparation of related documents to cover the issues discussed in the workshops
2. Collection of Q&A and discussed issues in the workshops from stakeholders of power, industries, transport, waste to energy, ministerial officers, NGO and private sectors
3. Summary report submitted to the policymakers in the related ministries

Barriers to Thailand's NAMAs have been identified in the stakeholder workshops, including barriers to support for CO_2 countermeasures (CMs). The identified CO_2 CMs under NAMAs show that Thailand and other developing countries have taken the responsibility to prepare voluntary mitigation in the low-carbon development plans toward helping, together with developed countries, to solve global climate issues. However, Thailand and other developing countries require capacity-building assistance in preparing CO_2 CMs under NAMAs. Further, developed countries need to provide support for developing countries in terms of capacity building, technology transfers of CO_2 CMs and financial support.

It was agreed on by the stakeholders that Thailand needs to analyse the policies of all involved agencies, the structure of the organisations involved, the existing energy policies related to GHG emission measures and the barriers to implementing GHG mitigation actions in renewable electricity generation and energy savings in industry, building and transport sectors.

The implementation of GHG mitigation actions within the renewable electricity generation sector and the energy savings in the industry and building sectors can be performed using two approaches—'project-based NAMAs' and 'sectoral-based NAMAs'. Both approaches have to be analysed for the amount of GHG reductions which will involve related organisations during implementation and MRV process. Thailand intends to set up an organisation or group responsible for the follow-up of NAMA implementation and NAMA MRV that will assume the role of 'Thailand NAMA coordinator'.

On the other hand, Thailand has yet to familiarise itself with the 'sectoral-based NAMAs' since most of the actions are nonmarket mechanisms where the structure and process of the GHG mitigation actions are set up by the government, e.g. adders and feed-in tariff scheme for electricity generation from renewable energy (solar, wind, hydro and biomass), which have different types of energy sources, NAMA methodology and NAMA MRV measures. The sectoral-based NAMAs can be adapted for use with other policies in the energy sector, e.g. measures to promote energy efficiency, whilst energy savings in the building and industrial sectors will have different NAMA methodologies and NAMA MRV processes.

6.10 Co-Benefits of Thailand's NAMAs

6.10.1 Energy Security Aspect

Co-benefits of GHG mitigation actions in Thailand NAMAs in terms of energy security have been assessed. The co-benefit aspects under consideration are of the following indicators: (1) Diversity of Primary Energy Demand (DoPED), (2) Net Energy Import Dependency (NEID), (3) Net Oil Import Dependency (NOID), (4) Net Gas Import Dependency (NGID) and (5) Non-Carbon-based Fuel Portfolio (NCFP), along with four co-benefits, which are (1) oil import intensity (OII), (2) gas import intensity (GII), (3) energy intensity (EI) and (4) carbon intensity (CI).

Results from analyses of energy security show that CO_2 countermeasure implementations under Thailand's NAMAs are able to increase national energy security since Thailand's NAMAs are related to the promotion of renewable electricity generation, which will also reduce the use of fossil fuels in electricity generation and increase the energy efficiency in the industrial and building sectors. In the case of a GHG reduction of 20 % in 2020, the indicators for DoPED and

NEID will be increased by 2.66 % and 10.56 %, respectively, whilst the indicators for NOID and NGID will be decreased by 3.65 % and 3.62 %, respectively, when compared to BAU for 2020. From the increase in renewable energy, the CO_2 emissions can be reduced, which will result in an increased NCEP indication of 18.99 %.

In addition, the indicators on OII and GII will be decreased by 8.14 % and 8.35 %, respectively. Furthermore, the indicators on energy intensity and carbon intensity in the case of a GHG reduction of 20 % in 2020 will be decreased by 17.9 % and 18.3 %, respectively, when compared to BAU.

Therefore, promoting the use of renewable energy can increase the energy security indices and GHG mitigation in Thailand. In general, Thailand still has many kinds of useful renewable energy resources which have high potential to be utilised because Thailand is an oil-importing country, so it is important that the government give more attention to promote the renewable energy resources to replace the imported fossil fuels (Limmeechokchai et al. 2014).

6.10.2 Environmental Aspect

Thailand NAMA implementation will directly result in decreased fossil fuel consumption. Consequently, a large amount of GHG emissions will be mitigated. In addition, other gases from combustion of fossil fuels will be mitigated as well. Transport NAMA actions will directly improve local and city air quality. It was also found that Thailand's NAMA actions will result in not only decreased CO_2 emissions, but also decreased CO, NO_x and SO_2 emissions.

6.10.3 Economic Aspect

The GHG countermeasures in Thailand's NAMAs have been assessed in terms of macroeconomic effects by using the input-output table. Results show that investment from the economic sector in Thailand's renewable energy power plants can be increased, which will cause an increase in the domestic production when compared to the 2020 BAU scenario. It also means an increase in investment for biomass and hydropower plant power generation, despite the fact that domestic production will be decreased due to solar and wind power projects. However, decreasing GDP in 2020 will come from investment of the private and public consumption sectors due to imported commodities resulting from the import of renewable technologies such as solar PV and wind turbines. Employment in the country will increase in 2020 as a result of increased activity in electricity production from domestic biomass.

Results of macroeconomic analyses based on increased energy efficiency of motors and lighting in the industrial sector show no economic stimulus, which will decrease the GDP by 1.25 billion Baht, increase the imported commodities by 1.19

billion Baht and increase the investment of the private and public consumption sector in implementing new motors by 0.83 billion Baht. However, the energy efficiency measures will help decrease the import of fossil fuels and gain more benefit when compared to the decreased GDP and also increase Thailand's energy security.

The analysis covering increased energy efficiency of cooling systems via installation of insulators for building envelopes and of lighting systems via upgraded light bulbs in the building sector shows that there is no increase in economic stimulus, which will decrease the GDP by 18.59 billion Baht, decrease the value of exports by 6.84 billion Baht and decrease the value of imports by 0.49 billion Baht. Moreover, the increased energy efficiency will decrease the private and public consumption sectors by 6.67 billion Baht and domestic production by 18.59 billion Baht. This means Thailand can reduce the amount of imported fossil fuels, which will outweigh the reduced GDP. This also increases Thailand's energy security.

6.10.4 Social Aspect

It is evident that the EE NAMA actions in the residential and commercial building sector will result in reduced energy cost for households and building owners. The social aspect of Thailand's NAMA actions has been assessed as savings per household and saving per unit of electricity consumption of buildings. In the NAMA case of a 20 % CO_2 reduction in 2020, it was found that the annual electricity bill saving per household will be around 60 USD. This social aspect of co-benefits shows the positive impact of Thailand NAMA actions, and finally it helps readily adopt EE NAMA actions.

6.11 Layout of Roadmap to Thailand NAMA 2020

To achieve a GHG reduction of 7–20 %, as Thailand communicated in its mitigation pledge to UNFCCC in Lima COP20, the Office of Natural Resources and Environmental Planning and Policy (ONEP) under MONRE proposed a roadmap to Thailand's 2020 NAMAs (see Fig. 6.9). This roadmap includes both domestically and internationally supported NAMAs (ONEP 2014). If GHG countermeasures implemented under Thailand's NAMA roadmap during 2014–2020 successfully clear the MRV processes, the higher GHG mitigation target of 20 % in 2020 will be achieved easily. However, the NAMAs still need support in terms of capacity building, financing, technology transfer and removal of EE barriers (Asayama and Limmeechokchai 2014).

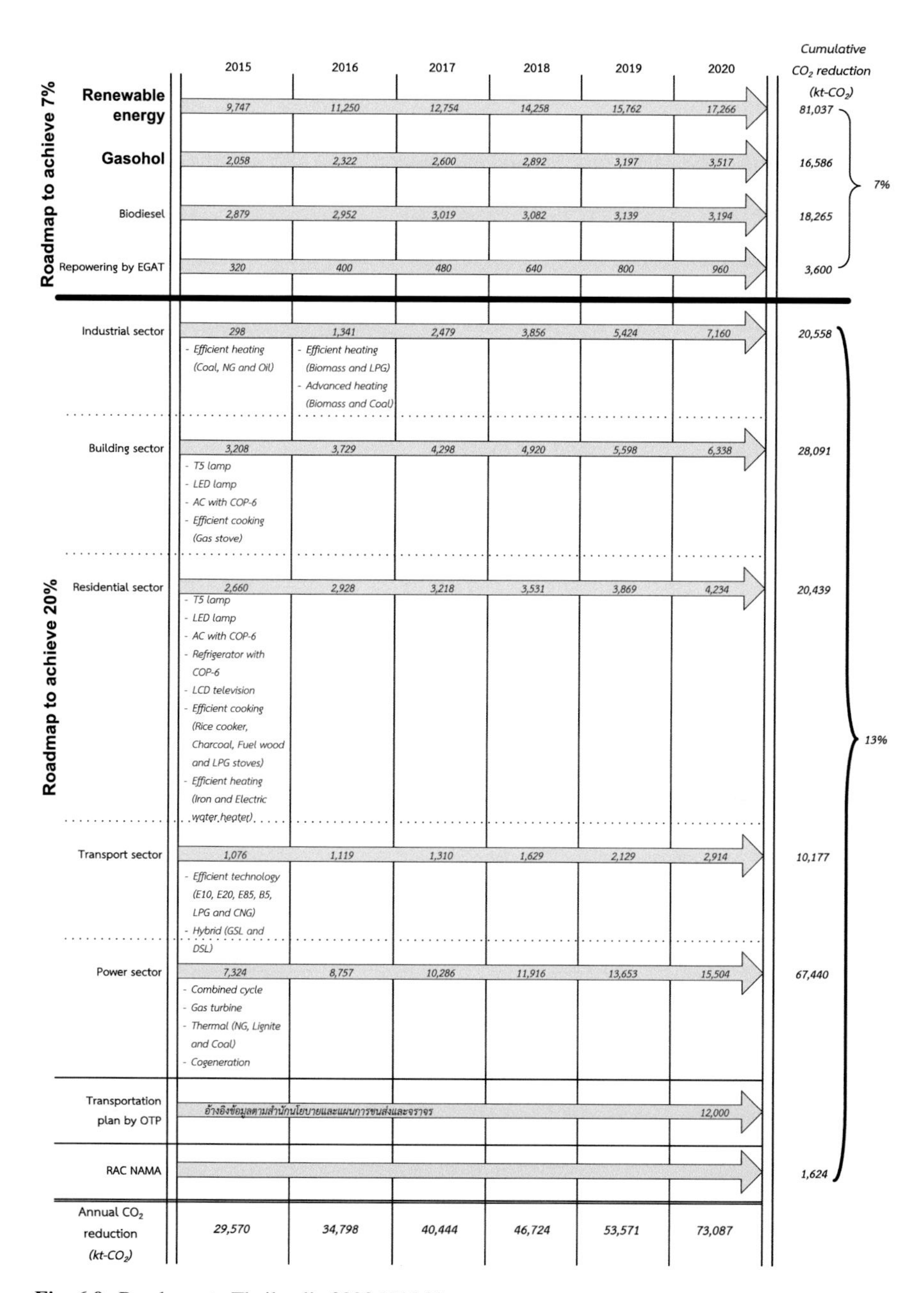

Fig. 6.9 Roadmap to Thailand's 2020 NAMAs

6.12 MRV of Thailand's NAMAs: The Road to Success

Thailand has prepared a NAMA roadmap in relation to the estimated GHG emission reductions. The MRV process under Thailand's NAMA actions complies with the UNFCCC process; thus, it will not cause more barriers to the NAMA MRV or the procedure according to the mitigation actions for both project-based NAMAs and policy-based NAMAs. Before pushing ahead with the projects, meetings with stakeholders must be held. These meetings need to be arranged by the agencies responsible for the MRV process, and MRV guidelines for each project need setting. This can be done by the working group and coordinator from the department of energy and climate change and by the working groups on GHG mitigation in the energy sector from the Ministry of Energy, together with the working groups from the Ministry of Energy and MONRE.

It was found that strategies for reduced GHG emissions from the use of renewable energy for electricity generation, increased energy efficiency in buildings and energy savings in industries are high priorities under Thailand's NAMA actions. However, for the NAMA actions to be achieved successfully, the organisations and the MRV process have to be developed at the same time as the NAMA action.

For the RE NAMA actions, the MRV process requires cooperation among ONEP, TGO, Ministry of Energy (Department of Alternative Energy Development and Efficiency), Energy Regulatory Commission (ERC) and Energy Policy and Planning Office (EPPO) (see Fig. 6.10a, b). These agencies will be responsible for NAMA MRV so that NAMA actions will be achieved to meet the targets of the AEDP plan and the targets for GHG emission reductions following the NAMA actions.

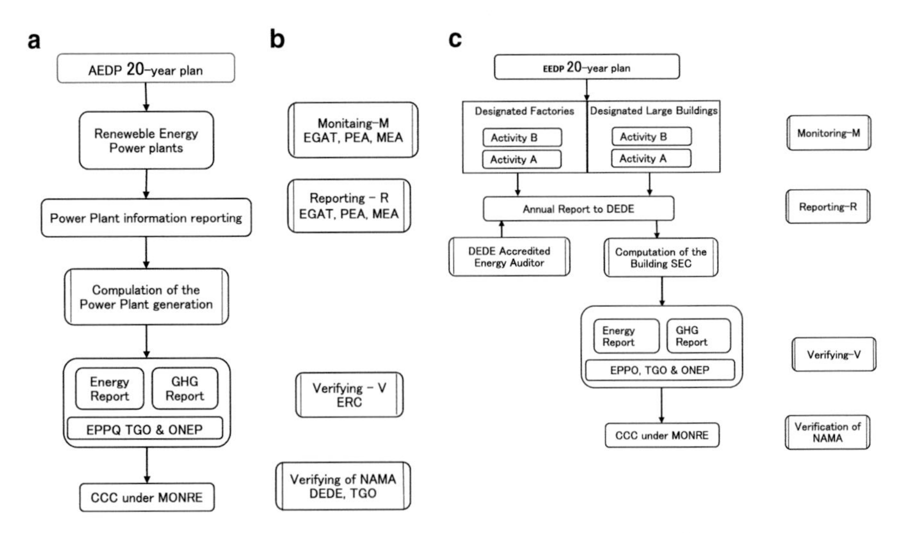

Fig. 6.10 MRV processes of domestically supported NAMAs (**a**) MRV of RE power generation (**b**) MRV of substitution of biofuels for fossils (**c**) MRV of energy efficiency in buildings and industries

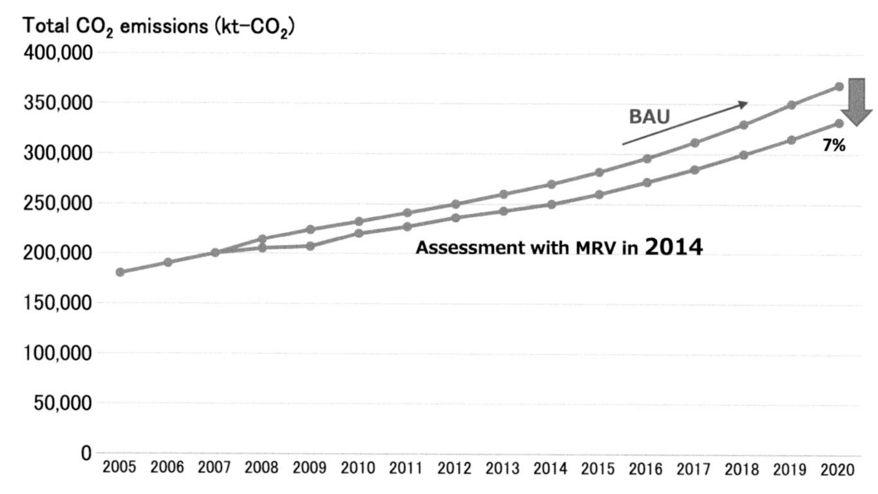

Fig. 6.11 Success of MRV process for domestically supported NAMAs

For the EE NAMAs, the actions are related to energy efficiency in industry and buildings. The MRV process requires cooperation among TGO and the Ministry of Energy (Department of Alternative Energy Development and Efficiency) (see Fig. 6.10c). These agencies will be responsible for NAMA MRV so that NAMA actions will accomplish the targets in energy conservation in industry and buildings and the targets for GHG emission reductions.

Finally, in October 2014, the ONEP revealed roadmap to Thailand's NAMAs. It shows that the domestic MRV processes of both RE and EE NAMAs already achieve a CO_2 reduction of 7 % (Fig. 6.11).

References

AEDP (2012) Department of Alternative Energy Development and Efficiency (DEDE) (2012) The Renewable and Alternative Energy Development Plan for 25 Percent in 10 Years (AEDP 2012. published by Ministry of Energy, Bangkok, Thailand Available: http://www4.dede.go.th/dede/images/stories/pdf/dede_aedp_2012

Asayama Y, Limmeechokchai B (2014) Policies and measures to remove energy efficiency barriers in Thai Buildings toward NAMAs. In: International conference and utility exhibition 2014 on green energy for sustainable development (ICUE 2014), Pattaya, Thailand, 19–21 March 2014

Department of Alternative Energy Development and Efficiency (DEDE) (2009) Building energy code. Ministry of Energy. Available: http://www.dede.go.th/dede/images/stories/energysaving12/15_2552.pdf

Department of Alternative Energy Development and Efficiency (DEDE) (2010) Annual report: Thailand energy situation 2010. Ministry of Energy. Available: http://www.dede.go.th/dede/index.php?option=com_content&view=article&id=1841&Itemid=318

Department of Alternative Energy Development and Efficiency (DEDE) (2012) The renewable and alternative energy development plan for 25 Percent in 10 Years (AEDP 2012–2021). Ministry of Energy. Available: http://www.dede.go.th/dede/images/stories/dede_aedp_2012_2021.pdf

Electricity Generating Authority of Thailand (EGAT) (2010) Thailand Power Development Plan (PDP2010: Revision3), published by EGAT, Bangkok, Thailand Available: http://www.egat.co.th/en/images/about-egat/PDP2010-Rev3-Eng.pdf

Energy Policy and Planning Office (EPPO) (2011) Thailand 20-year energy efficiency development plan (2011–2030). Ministry of Energy. Available: http://www.eppo.go.th/encon/ee-20yrs/EEDP_Eng.pdf

Kainuma M, Matsuoka Y, Morita T (2003) Climate policy assessment: Asia-Pacific integrated modelling, vol II. Springer, Tokyo

Limmeechokchai B, Winyuchakrit P (2014) Development of Thailand's Nationally Appropriate Mitigation Actions (NAMAs) for low carbon society: energy security and co-benefit aspects. In: International Conference and Utility Exhibition 2014 on Green Energy for Sustainable Development (ICUE 2014), Pattaya, Thailand, 19–21 March 2014

Limmeechokchai B, Winyuchakrit P, Promjiraprawat K, Selvakkumaran S, Sritong N, Chunark P (2013) The development of mitigation pledge and Nationally Appropriate Mitigation Actions (NAMAs) in energy sectors: power generation, industry and designated buildings. Thammasat University Research and Consultancy Institute, Pathumthani

Office of Natural resources and Environment Planning and Policy (ONEP) (2014) Roadmap to Thailand NAMA 2020, published by ONEP, Bangkok, Thailand, December 2014

ONEP (2011) Thailand second national communication. Office of Natural Resources and Environmental Policy and Planning (ONEP), Ministry of Natural Resources and Environment, Bangkok, Thailand. Available: https://unfccc.int/files/national_reports/nonannex_i_natcom/submitted_natcom/application/pdf/snc_thailand.pdf

Thailand Greenhouse gas organization Management (TGO) (2014) Status of carbon credits in Thailand, published by TGO, Bangkok, Thailand, December 2014

Chapter 7
'Science to Action' of the Sustainable Low Carbon City-region

Lessons Learnt from Iskandar Malaysia

Chin Siong Ho, Loon Wai Chau, Bor Tsong Teh, Yuzuru Matsuoka, and Kei Gomi

Abstract This paper outlines the lessons learnt through the multidisciplinary 'Science-to-Action' approach to formulating, mainstreaming and implementing the *Low Carbon Society Blueprint for Iskandar Malaysia 2025* (LCSBP-IM2025). Iskandar Malaysia (IM) is a rapidly developing urban region in southern Peninsular Malaysia that was institutionalised in 2006 with a view to spurring Malaysia's economic growth up to 2025. In pursuing rapid economic growth to become a developed, high-income nation by 2020, Malaysia is conscious of its global responsibility in environmental protection and global climate change mitigation, hence the country's commitment to reducing its carbon emission intensity of GDP by up to 40 % by 2020 based on the 2005 level. Being a premier economic region in Malaysia, IM seeks to develop a low carbon society (LCS) and lead the way to cutting its carbon emission intensity by up to 58 % by 2025 based on the 2005 level through the implementation of the LCSBP-IM2025.

The LCSBP-IM2025 is the outcome of an internationally funded joint research under the SATREPS programme that brings together Universiti Teknologi Malaysia (UTM), Kyoto University, Japan's National Institute for Environmental Studies (NIES), Okayama University and the Iskandar Regional Development Authority (IRDA), in a unique 'academia-policymaker' partnership, towards crafting an LCS pathway to guide and sustainably manage the projected rapid development in IM up to 2025. To that end, a multidisciplinary research team that comprises the above research institutions and IRDA, led by UTM, has been set up. A methodology has

C.S. Ho • L.W. Chau (✉) • B.T. Teh
UTM-Low Carbon Asia Research Centre, Faculty of Built Environment, Universiti Teknologi Malaysia, Johor Bahru, Malaysia
e-mail: lwchau@utm.my

Y. Matsuoka
Graduate School of Engineering, Kyoto University, Kyoto, Japan

K. Gomi
Center for Social and Environmental Systems Research, National Institute for Environmental Studies, Tsukuba, Japan

S. Nishioka (ed.), *Enabling Asia to Stabilise the Climate*,
DOI 10.1007/978-981-287-826-7_7

been developed to formulate IM's future LCS scenarios, propose LCS actions to achieve the LCS scenarios, quantify the GHG emission reduction potential of the proposed LCS actions and continuously engage local stakeholders in a series of focus group discussions (FGDs).

The project has been a great success from its official commencement in July 2011, which saw the LCSBP-IM2025 being launched at UNFCCC's COP 18 in Doha in November 2012 and officially endorsed by the Malaysian Prime Minister in December the same year. In November 2013, the *Iskandar Malaysia Actions for a Low Carbon Future* was launched, outlining ten priority projects selected from the LCSBP-IM2025's 281 programmes for implementation in IM in 2013–2015; the projects are now at various stages of implementation, yielding real impacts on IM's progression towards its LCS goal.

The project offers valuable lessons especially in advancing *scientific research* on LCS into *policymaking* and, importantly, into *real actions* (hence, Science to Action). These include the importance of having strong highest-level government support, aligning LCS actions to higher-level development priorities, taking policymakers on-board the research team, continuously actively engaging local communities and stakeholders through FGDs and overcoming science-policy and disciplinary gaps that emerged. What is clearly evidenced by the LCSBP-IM2025's success is that developing countries, with good synergy between highly committed local research institutions and policymakers, subject to adequate international funding and technological assistance from developed nations, are capable of crafting and putting in place implementable LCS policies that eventually contribute to mitigating global climate change through real cuts in GHG emissions while still achieving a desired level of economic growth.

Keywords Low carbon society (LCS) • Science to Action (S2A) • Iskandar Malaysia • Low carbon society blueprint • Academia-policymaker partnership • Extended Snapshot Tool (ExSS) • GHG emission mitigation • Green growth • Urban region • Lessons learnt

Key Messages to Policymakers

- Low carbon society is the way forward to strong, sustainable cities and regions.
- Internationally funded joint research on LCS is essential to developing countries.
- Good scientific research is cornerstone to effective implementation of LCS policies.
- Policies supported by science are effective for realising GHG emission reduction.
- Highest-level government support greatly expedites LCS science to LCS actions.

7.1 Introduction

Malaysia, like most other rapidly urbanising ASEAN countries, though not a significant source of emissions of greenhouse gases (GHG), has taken actions to address climate change through various environmental, economic and social initiatives over the years. In 2009, Malaysia voluntarily set a target for GHG reduction of up to 40 % in terms of energy intensity of GDP by 2020 compared to 2005 levels. Following that, a series of national-level key policies aiming at guiding the nation towards addressing climate change holistically, ensuring climate-resilient development, developing a low carbon economy and promoting green technology have been formulated. These include the *National Policy on Climate Change* (MNRE 2009), *National Green Technology Policy* (KeTTHA 2009a), *National Renewable Energy Policy and Action Plan* (KeTTHA 2009b) and the *Green Neighbourhood Planning Guideline* (JPBDSM 2012), among others. These are important in providing a framework for achieving Malaysia's broader sustainable development goals, while the country elevates itself to become a high-income nation by 2020 (PEMANDU 2010). At the national level, the Malaysian Government is positioning the ecosystem, value system and supply chain to create a vibrant low-carbon economy. Apart from national mitigation and adaptation strategies for addressing the impact of climate change, there is a need to also look into regional and local resilient policies to reduce GHG emissions, especially in major cities and economic development corridors involving many urban conurbations. It is indeed at the regional and local levels that climate change policies may be operationalised and see their effects.

The International Energy Agency estimates that urban areas currently account for two thirds of the world's energy-related GHG emissions, and this is expected to rise to about 74 % by 2030 (World Bank 2010). Cities especially in developing countries with rapid population growth and economic development are consuming vast natural resources, generating enormous amounts of wastes and emitting large volumes of GHGs. Despite the fact that cities are the main carbon emission contributors, experts largely agree that cities nonetheless offer the greatest opportunity for mitigating climate change. City-based climate change policies are proven to be effective and efficient, feasible and relatively easy to deliver as compared to national climate change policies. Many cities, predominantly in developed countries, have established action plans and road maps to tackle climate change issues. However, difficult challenges lie ahead of cities in developing and transition nations in Asian regions, including Malaysian cities, where urban population is high and growing fast, economic growth is rapid and general awareness of climate change is relatively low; there appears to be an observable lack of knowledge, experience and urgency in mitigating climate change at the city and regional levels.

In line with the Malaysian Government's objectives to strengthen economic competitiveness and improve quality of life, and its aspiration for promoting green economic growth and greater sustainability, Iskandar Malaysia (IM), a rapidly developing economic corridor established in 2006, sets out to be the first urban region in Malaysia to formulate and implement a city-regional level climate change action plan – the *Low Carbon Society Blueprint for Iskandar Malaysia 2025* (LCSBP-IM2025). Optimistically, the LCSBP-IM2025, being perhaps among the first few city-regional level climate change action plans in developing countries, does not only benefit IM in laying out a clear sustainable development pathway for the urban region but also other Malaysian and Asian cities and regions through the sharing and dissemination of good practice and experiences gained in drawing up the Blueprint for implementation. The purpose of this paper is to outline the experiences gained and lessons learnt through the multidisciplinary 'Science to Action' (science to policy to implementation) approach to drawing up and mainstreaming the LCSBP-IM2025 for implementation in IM.

7.1.1 About Low Carbon Society Blueprint for Iskandar Malaysia 2025

The *Low Carbon Society Blueprint for Iskandar Malaysia 2025* is a written document that presents comprehensive climate change mitigation policies and detailed strategies to guide the development of Iskandar Malaysia towards becoming 'a strong and sustainable metropolis of international standing' in 2025, in line with the urban region's development vision. The LCSBP-IM2025 incorporates various related national policies, the *Comprehensive Development Plan for South Johor Economic Region 2006–2025* (CDP) (Khazanah Nasional 2006) and 24 Iskandar Malaysia blueprints towards transforming IM into a sustainable, low carbon metropolis that is built on solid economic foundations (for more details on the policy context and framework of the Blueprint, see Sect. 7.2). The LCSBP-IM2025 provides and explains technical details of carbon mitigation options (with specific measures and programmes) for implementation in IM. It is aimed at coordinating and guiding the implementation of a total of 281 programmes organised under 12 low carbon society (LCS) policy actions in IM in order to lead the urban region towards achieving a targeted 50 % reduction in GHG emission intensity of GDP by 2025 based on the 2005 level.

7.1.2 Low Carbon Society (LCS)

The concept of low carbon society (LCS) is the fundamental philosophy that underpins the formulation of the LCSBP-IM2025. LCS is an emerging theory and is defined as (Skea and Nishioka 2008, p. S6):

> A society that takes actions that are compatible with the principles of sustainable development, ensuring that the development needs of all groups within the society are met.

> A society that makes an equitable contribution towards the global effort to stabilise atmospheric concentrations of carbon dioxide and other greenhouse gases (GHGs) at a level that will avoid dangerous climate change through deep cuts in global emissions.
>
> A society that demonstrates high levels of energy efficiency and uses low carbon energy resources and production technologies.
>
> A society that adopts patterns of consumption and behaviour that are consistent with low levels of GHGs emission.

The ideology of LCS emphasises 'people' – the society – as the source of, and at the same time, solution to, climate change. It highlights existing human activities as the main contributors to global GHG emissions and therefore calls for efforts of the current society in all sectors to shift their mass consumption behaviour and lifestyle to a new consumption pattern that poses less harm to the environment. Low carbon society is a new society that consumes relatively low amounts of resources (raw materials, energy and water) in minimising GHG emissions to avoid adverse effects of climate change. Despite the fact that the concept stresses on social reform for better environmental system, it does not compromise the attainment of robust economic growth and maintenance of high quality of life. In this light, there are two fundamental aspects of LCS in leading societies towards reducing GHG emissions:

1. 'Decoupling' of economic activities, urban growth and urban transportation from intense resource and energy consumption and GHG emissions towards minimising the environmental impacts of increased economic activities and transportation (see Li 2011; UNEP 2011, 2014)
2. Exploration for attainment of potential social, environmental and economic 'co-benefits' arising out of climate change policies, which have been found to be highly pertinent to effective implementation at the local city level and to getting greater political acceptance of the policies (see de Oliveira et al. 2013; Seto et al. 2014)

In realising LCS, various 'soft' and 'hard' infrastructure developments/improvements are needed to encourage communities to change their preferences and behaviours to the practice of green lifestyles. 'Soft' infrastructure includes intangible elements that comprise of awareness, education, governance, institutions, legislation and finance. On the other hand, 'hard' infrastructure refers to physical elements that include the urban form, land use structure, transportation system, technology, building design and utilities (see Fig. 7.1). Specific strategies for low carbon society transformation for one city will be different from another city with respect to their geographic, economic, political and sociocultural contexts.

7.1.3 Iskandar Malaysia (IM) in Brief

Iskandar Malaysia, previously known as the South Johor Economic Region (SJER) and the Iskandar Development Region (IDR), is a visionary economic region in the

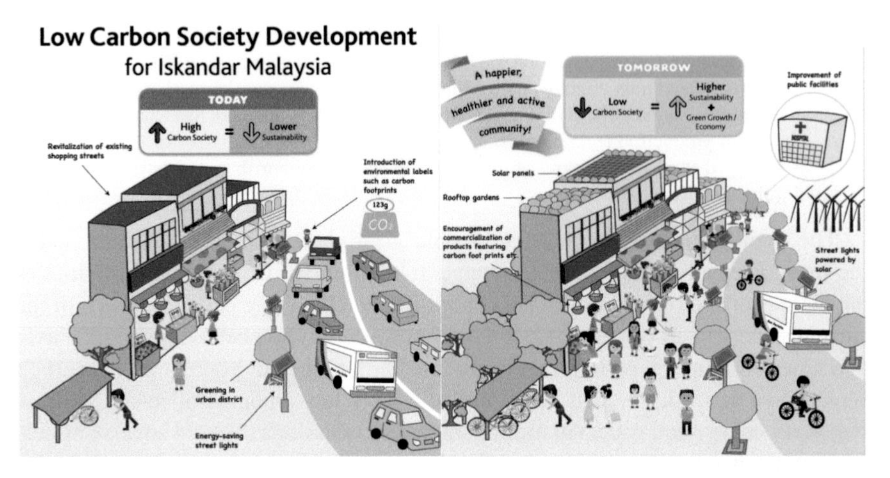

Fig. 7.1 Schematic representations of social-environmental-economic transformations involving changes in lifestyle and technology that will bring about a low carbon society in Iskandar Malaysia (Source: IRDA 2014)

southern tip of Peninsular Malaysia. The region with the size of 221,634 ha (2,216.3 km^2) was established in 2006 as one of the catalyst development corridors to spur growth of the Malaysian economy into the first quarter of the twenty-first century. In the macro-regional context, IM is strategically located at the southernmost tip of Mainland Asia to tap on a vast and burgeoning market of about 0.8 billion people within a 6-hour flight radius (Fig. 7.2).

Envisioned to be 'a strong and sustainable metropolis of international standing' and set to become an integrated global node that synergises with growth of the Global City State of Singapore and the Riau-Batam Region of Indonesia, it has been projected that IM will be sustained by a rapid annual gross domestic product (GDP) growth of 7–8 % that will almost quadruple the GDP of the urban region to MYR141.4 billion in 2025. The urban region is expected to experience a concomitant rapid population growth, with Iskandarians more than doubling from 1.35 million in 2005 to about 3 million by 2025 (Khazanah Nasional 2006).

As shown in Fig. 7.2, five flagship zones have been established as the main economic growth centres with their respective niche sectors in IM. These flagship zones have been envisaged to both further strengthen and value-add to the urban region's existing economic clusters as well as to diversify and develop targeted strategic growth sectors.

In terms of local administration, IM covers the entire Districts of Johor Bahru and Kulai and three subdistricts of Pontian. The administrative jurisdiction of the urban region falls under five local authorities (which are also the respective local planning authorities (LPAs) for their areas), namely, the Johor Bahru City Council (MBJB), Johor Bahru Tengah Municipal Council (MPJBT), Pasir Gudang Municipal Council (MPPG), Kulai Municipal Council (MPKu) and Pontian District Council (MDP) (see Fig. 7.3). In addition to traditional state and local

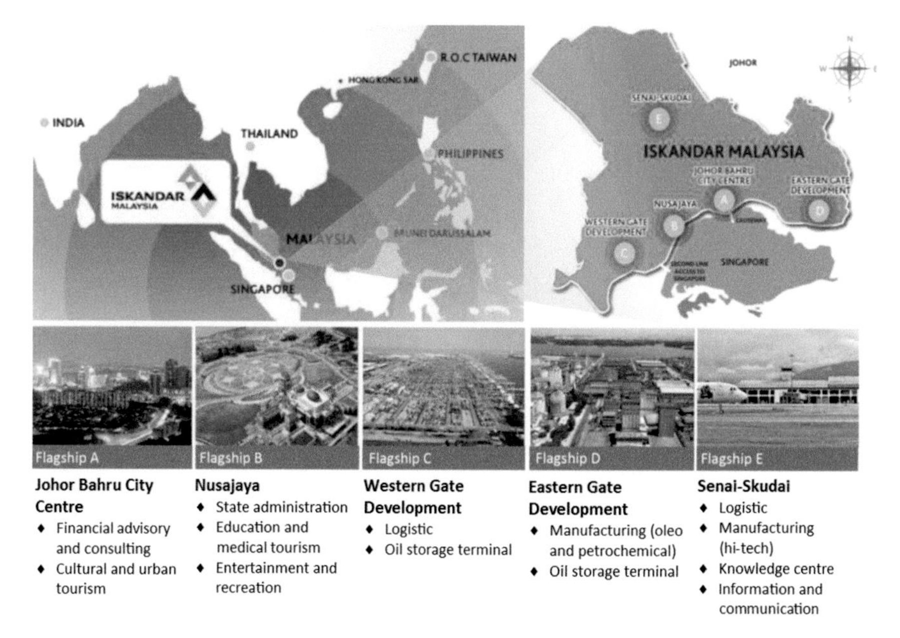

Fig. 7.2 Geographic location of Iskandar Malaysia within the Southeast Asian region and the five flagship zones in Iskandar Malaysia (Source: Adapted from http://www.iskandarmalaysia.com.my/)

administrative institutions, towards expediting and ensuring effective implementation and materialisation of development plans and policies in IM, the Iskandar Regional Development Authority (IRDA) has been set up under the Iskandar Regional Development Authority Act, 2007 (Act 664). Under this Parliamentary Act, IRDA which is co-chaired by the Malaysian Prime Minister and Johor State *Menteri Besar* (literally the 'Chief Minister') holds the statutory functions of planning, promoting, coordinating and facilitating development and investments in IM.

As Iskandar Malaysia undergoes very rapid physical-spatial development and economic growth, it becomes highly essential that the social and environmental impacts of its rapid expansion and economic growth are mitigated, guided by a holistic LCS blueprint that will fit into the existing development planning and institutional framework presently at work in the urban region.

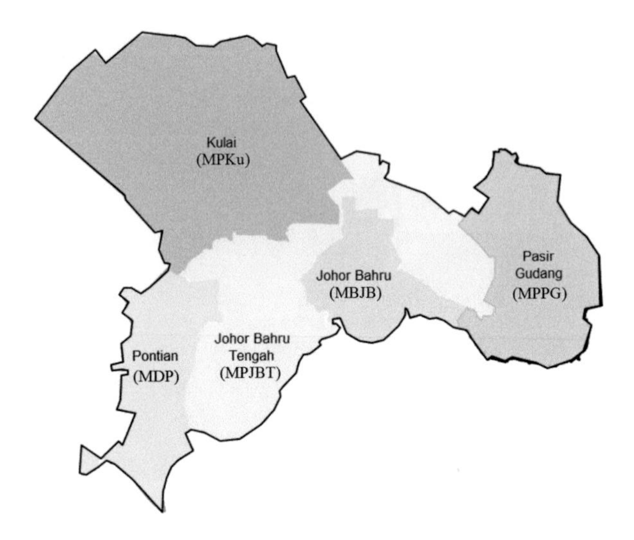

Fig. 7.3 Municipal jurisdictions within the Iskandar Malaysia economic corridor (Source: Adapted from Khazanah Nasional 2006)

7.2 Integrating Low Carbon Society Blueprint into Existing Policy Framework

Since its inception in 2006, development in Iskandar Malaysia has been governed by various policies, plans and guidelines at the national, state and local levels. Specifically, IM has a statutory Comprehensive Development Plan (CDP) that is provided for under the Iskandar Regional Development Authority Act, 2007 (Act 664) and a series of 24 blueprints covering various development aspects of the urban region (see Fig. 7.4, column 2); the blueprints gain statutory status by means of adoption by the Johor State Planning Committee (SPC) under the Town and Country Planning Act, 1976 (Act 172). The main function of the CDP and blueprints is to provide a development coordination framework by which all government entities within Iskandar Malaysia are to legally abide under Act 664.

At the same time, IM is also home to five local planning authorities (LPAs) that hold the traditional statutory role of planning and regulating development and use of land within their administrative areas under Act 172. The LPAs come under the *Johor Bahru District and Kulai District Local Plan 2020*, which is the statutory plan provided for under Act 172 for guiding and regulating land use and development in the Johor Bahru and Kulai Districts (which jointly cover most of Iskandar Malaysia) (Fig. 7.4, column 3). The Local Plan is required by law to take cognisance of and provide clear spatial articulation to higher-level development policies, including the *Johor State Structure Plan 2020*, the *National Physical Plan-2* as well as other general development policies (Fig. 7.4, column 4). Most LPAs also enact their respective development policies and planning guidelines that have to be in line with the Local Plan. However, reducing energy and carbon emission intensity of rapid growth has to date not been an agenda of these plans and policies.

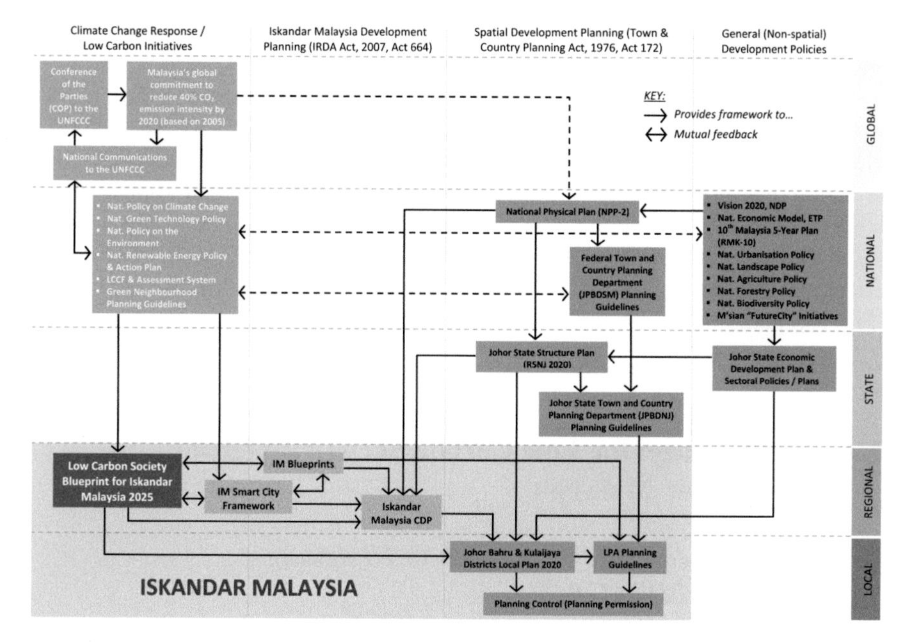

Fig. 7.4 Policy context of the LCSBP-IM2005 which serves as the critical link between global and national climate change initiatives and policies and local development planning policies and regulation mechanisms (Source: UTM-Low Carbon Asia Research Centre 2013a, p. 3–4)

Since the honourable Prime Minister of Malaysia made the pledge of voluntary reduction of the country's carbon emission intensity at COP 15 in 2009, a series of national-level climate change responses and low-carbon initiatives have emerged in the forms of policies, framework and guidelines (Fig. 7.4, column 1). However, these policies and guidelines have yet to find their way into the lower-level development policies, plans and guidelines that are more effective and detailed in guiding and regulating physical-spatial development but are hitherto largely 'carbon blind'.

Being a premier economic corridor in Malaysia, it is only appropriate that IM leads the way in contributing to honouring the country's pledge to reduce its carbon emission intensity by 40 % (based on 2005 emission levels) by 2020. It is in this light that the *Low Carbon Society Blueprint for Iskandar Malaysia 2025* is formulated to provide the crucial policy link between the country's global and national climate change responses (Fig. 7.4, column 1) and Iskandar Malaysia's regional- and local-level development plans and policies. To that end, the Blueprint sets a GHG emission intensity reduction target of 50 % by 2025, based on the 2005 emission level. The target would be achieved through implementing 12 LCS actions set under three main themes: *Green Economy*, *Green Community* and *Green Environment*. The Blueprint also takes special cognisance of the recently launched Iskandar Malaysia Smart City Framework that sets out the general characteristics of

IM as a smart city, which include elements of reducing carbon emission and emphasis on development of ICT infrastructure.

Once adopted by the SPC, the Blueprint shall provide a statutory policy framework for the CDP, which is currently under review, and serve as the 'umbrella blueprint' for the existing 24 IM blueprints which need to be progressively revised to incorporate relevant LCS policies and strategies. As required under Act 664, these would subsequently trickle down to the *Johor Bahru District and Kulai District Local Plan 2020* and various LPA planning guidelines and take effect through the granting of planning permissions to future developments in IM (Fig. 7.4, pink box).

7.3 Policy Design for Low Carbon Society Blueprint in Iskandar Malaysia

7.3.1 Science-to-Policy Approach to Designing the LCSBP-IM2005

The LCSBP-IM2025 is developed from a unique 'academia-policymaker' partnership with the involvement of various stakeholders (local communities; NGOs; businesses and industries; Federal, State and local authorities) along the way. The application of well-tested scientific modelling to inform LCS policies and the promotion of green technologies and industries in the Blueprint towards achieving industrial growth and social well-being and transforming governance are in line with the recent 'Science to Action' (S2A) initiatives championed by the Malaysian Prime Minister (New Partnership for Climate Resilient Development 2014).

Through S2A, the government aims to intensify the application of science and technology as a key pillar of the nation's development and the *rakyat's* (people's) well-being. In the context of the LCSBP-IM2025, the application of science and technology is in the area of climate change mitigation, environmental protection and urban planning for urban-regional development. It involves the joint efforts between four research institutions from Malaysia (Universiti Teknologi Malaysia) and Japan (Kyoto University, National Institute for Environmental Studies (NIES) and Okayama University) and also a regional development authority (Iskandar Regional Development Authority) that is responsible for coordinating and enabling development in the Iskandar Malaysia region (see Fig. 7.5).

The overall process as shown in Fig. 7.5 begins with the usual information gathering, analysis and contextual appraisal of current development, carbon emission and policy scenarios in Iskandar Malaysia. This informs the setting of IM's LCS goals and carbon emission reduction target in 2025. These then feed into an iterative process of formulating policy actions, measures and programmes and testing them via the *Asia-Pacific Integrated Model* (AIM) against the achievement of set goals and targets. The AIM is a suite of scenario-based quantitative

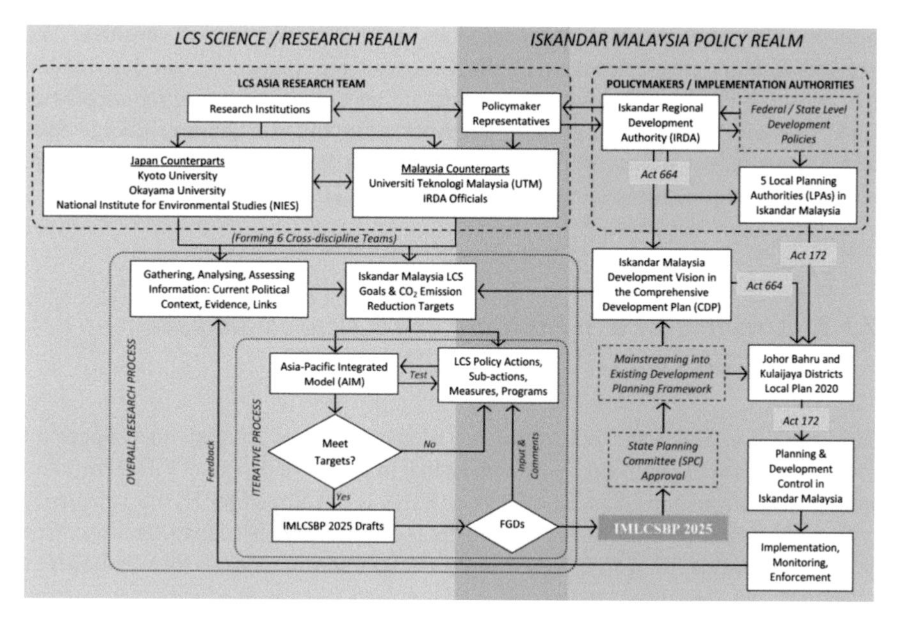

Fig. 7.5 The science/research-policymaking model that emerged from the formulation of the LCSBP-IM2025 and mainstreaming of the Blueprint into the existing development planning institutional framework (Source: UTM-Low Carbon Asia Research Centre 2013a, p. 0–6)

accounting tools that are able to both forecast multiple-scenario GHG emissions and then 'backcast' based on the selected GHG emission reduction scenario to guide timely implementation of policies and allocation of resources in order to achieve the emission reduction target. The AIM has been widely used in many countries and is recognised by the Intergovernmental Panel on Climate Change (IPCC), an international leading body for the assessment of climate change. The main tool used to forecast GHG emissions of different scenarios in IM – the Extended Snapshot tool (ExSS) – will be explained in more detail below (see Sects. 7.3.2 and 7.3.3).

Stakeholder participation is built into the process at this stage through a series of focus group discussions (FGDs) where proposed actions, measures and programmes are scrutinised by stakeholders and their opinions are gathered and fed back into the policy formulation process. A total of five rounds (nine sessions) of FGDs have been held until the final draft of the Blueprint was ready for consideration for approval by the State Planning Committee (SPC) and subsequent mainstreaming (see UN-Habitat 2012) into the existing development planning framework for implementation.

The LCSBP-IM2025 is therefore formulated based on scientific and quantitative modelling that incorporates cyclical input and feedback from various stakeholders, resulting in practical and feasible LCS policies with improved public acceptance, corporate buy-in and eventual policymaker adoption and implementation of the policies. The Blueprint thus exemplifies effective sustainable development policymaking that is scientifically based and institutionally context sensitive. The holistic and integrated features of the Blueprint are shaped by six interrelated multidisciplinary expert groups from Malaysia and Japan, covering the aspects of Land Use and Scenario Integration; Transportation and Air Quality; Energy Systems; Sustainable Waste Management, Education and Consensus Building and regional development planning and governance (IRDA).

7.3.2 Creating LCS Scenarios – The Extended Snapshot (ExSS) Tool

This section explains the procedure and methodology of the Extended Snapshot tool (ExSS) in GHG emissions accounting that informs the design of GHG emission mitigation options for Iskandar Malaysia. ExSS is developed by Kyoto University and the National Institute for Environmental Studies (NIES), Japan, and was first launched in 2006 (Ali et al. 2013). It is a static accounting model with simultaneous equations and the ability to project consistent socio-economic variables, energy demand and supply and CO_2 emissions from energy consumption in a particular future year based on a set of future assumptions of development and energy technologies. The tool quantifies economic growth and changes in industrial structure; demography; changes of lifestyles in terms of consumption pattern and energy service demand; transport volume and structure; and low carbon measures that include energy-efficient devices and buildings, renewable energy, modal shift to public transport and fuel mix in power generation (see Gomi et al. 2010).

The methodology for creating LCS scenarios builds on the idea of 'backcasting', which begins with the setting of a desirable LCS goal followed by iterative explorations of possible options to achieve the goal using ExSS. Figure 7.6 summarises the overall process of the method which comprises seven steps.

1. Setting the framework

Framework of an LCS scenario includes a target area, a base year, target year(s), environmental targets and a number of scenarios. The base year provides the base scenario against which the target year scenario is compared. The target year should be far enough to realise the required change and yet near enough to capture with reasonable clarity the development vision and future scenarios in the target area. In the preparation of the LCSBP-IM2025, 2005 is selected as the base year, while the target year of IM's LCS scenario has been set as 2025. For the environmental target, CO_2 from energy use is opted for because it is expected to be a main source of GHG emissions in IM in 2025.

2. Descriptions of socio-economic assumptions

Before conducting the quantitative estimation, the qualitative future image of the target area's development is narrated. It is essentially an image of demography,

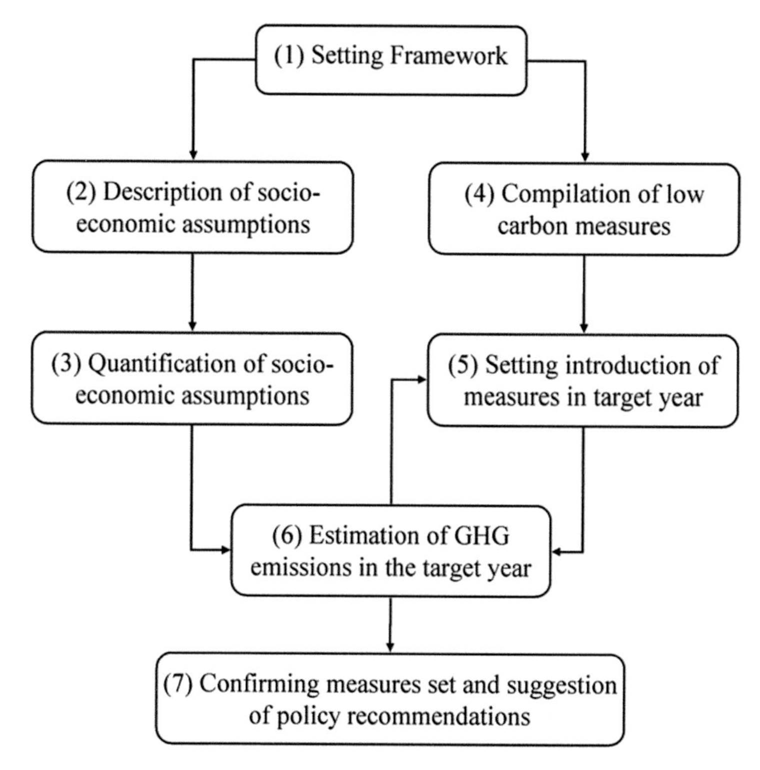

Fig. 7.6 Creating LCS scenarios, overall process (Source: Gomi et al. 2010, p. 4786)

lifestyle, economy and industry, land use, transportation, technology available, its diffusion level and so on. For the purpose of LCSBP-IM2025, Iskandar Malaysia's Comprehensive Development Plan (CDP) and various State and local official economic and development planning documents have the main sources on which future scenarios of IM are based.

3. Quantification of socio-economic assumptions

To provide 'snapshots' of estimated energy use and GHG emissions based on the future image of IM in Step (2), values of exogenous variables and parameters are set. These are then input into ExSS which then calculates various socio-economic indices of the target year, including population, GDP, output by industry, passenger and freight transport demand.

4. Compilation of low carbon measures

The next step involves the compilation of countermeasures (CM) which are expected to be available in the target year, for example, high-energy-efficiency devices, transport structure change such as public transport, use of renewable energy, energy-saving behaviour and carbon sink. Technical data are required to estimate their effects on reducing GHG emissions. For the purpose of the LCSBP-

IM2025, the technical data used have been based on those from a preceding study in Japan's Shiga Prefecture due to limited availability of IM-specific information and, importantly, similarity in the industrial structure and population size of the Shiga and IM regions.

5. Setting introduction of countermeasures

Technological parameters related to energy demand and CO_2 emissions, in short energy efficiency, are defined at this stage. Since there can be various portfolios of the measures, it is crucial that appropriate criteria are chosen, for example, cost minimisation, acceptance to stakeholders (through FGDs), realistic levels of technological development and their diffusion rates.

6. Estimation of GHG emission in the target year

Based on the socio-economic indices and assumptions of countermeasures' introduction set in Steps (3) and (5), GHG emissions are finally calculated using ExSS. If the resultant GHG emissions meet the preset reduction target, the correspondent combinations of countermeasures are selected for policy proposal in the next step. Otherwise, Step (5) will be repeated where countermeasures and technological parameters are reset until the GHG reduction target is achieved.

7. Proposal of policies

Policy set to introduce the countermeasures defined is proposed. Available policies depend on the context of the municipality, region or country which they are aimed at addressing. ExSS can calculate emission reduction potential for each countermeasure. Therefore, it can show the reduction potential of measures which especially need to be prioritised. It can also identify measures which have high reduction potential and are therefore important.

7.3.3 Structure of Extended Snapshot (ExSS) Tool

The Extended Snapshot tool is a key component of the aforementioned Asia-Pacific Integrated Model (AIM) developed by Kyoto University and NIES, Japan. It is a modelling tool to assess future energy consumption, power generation, technology diffusion, transportation, industrial outputs, residential and commercial activities and waste generation and GHG emissions, coupling with predetermined socio-economic, industrial and demographic scenarios in a particular future year (the target year).

Figure 7.7 shows the simplified internal working and data structure of the ExSS tool, which comprises four modules (driving forces, energy service demand, primary energy supply and GHG emissions) with input parameters, exogenous variables and variables between modules. ExSS is a system of simultaneous equations. Given a set of exogenous variables and parameters, solution is uniquely defined. In this simulation tool, only CO_2 emissions from energy consumption are calculated.

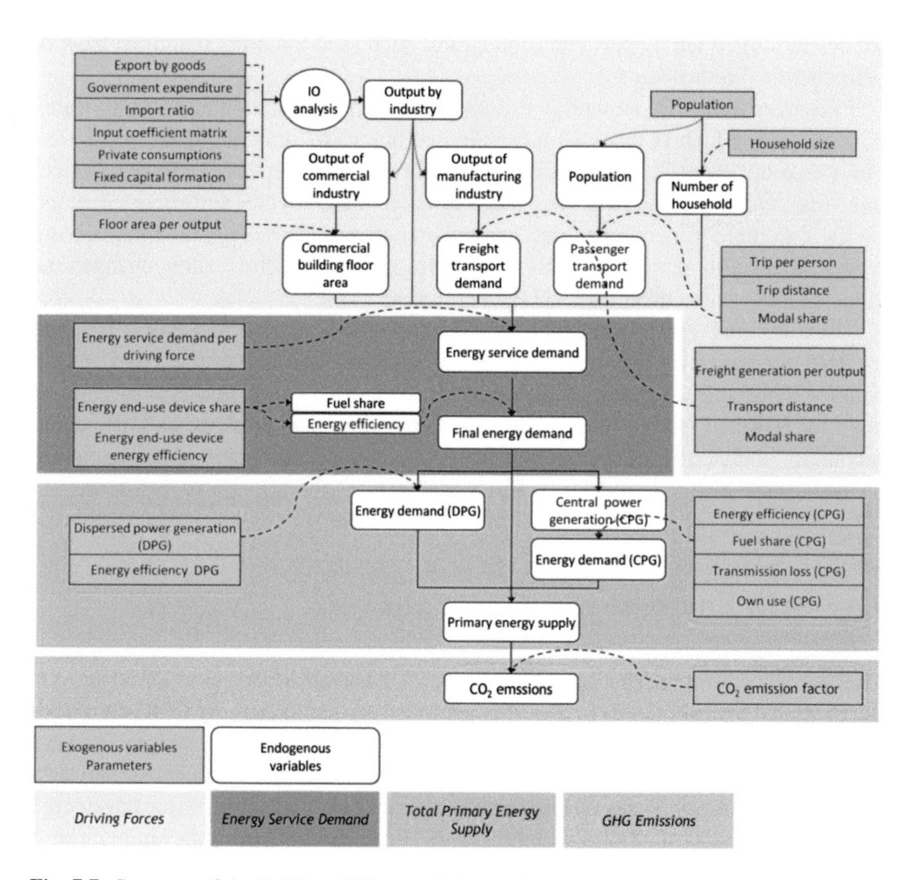

Fig. 7.7 Structure of the ExSS tool (Source: Adapted from Gomi et al. 2010)

In many LCS scenarios, exogenously fixed population data are used. However, people migrate more easily, when the target region is a relatively smaller area such as a state, district, city or town. Population is decided by demand from outside of the region, labour participation ratio, demographic composition and relationship of commuting with outside of the region.

To determine output of industries, the 'export-base' input-output approach is combined in line with the theory of regional economics. Industries producing export goods are called basic industries. Production of basic industries induces other industries, i.e. nonbasic industries, through demand of intermediate input and consumption of their employees. A number of workers must fulfil labour demand of those productions. Given assumptions of where those workers live and labour participation ratio, population living in the region is computed. This model enables us to consider viewpoints of regional economic development to estimate energy demand and CO_2 emissions. For future estimation, assumption of export value is especially important if future development of the target region is expected to (or desired to) be led by particular industries, such as automotive manufacturing or petrochemical industries.

Passenger transport demand is estimated from the population and freight transport demand, which is taken as a function of output by manufacturing industries. Floor area of commercial activities is determined from output of tertiary (service) industries. Other than driving force, activity level of each sector and energy demand by fuels are determined with three parameters: energy service demand per driving force, energy efficiency and fuel share. Diffusion of countermeasures changes the value of these parameters and so GHG emissions.

The estimated results of the future socio-economic indicators and energy demand in 2025 are based on the modelling of the socio-economic variables and energy balance table in 2025. Most of the socio-economic indicators and energy balance table for Iskandar Malaysia are obtained from official and published statistics and secondary sources. Assumptions are used where information for macroeconomic analysis is not available for the Iskandar Malaysia region (see Ho et al. 2010).

7.4 GHG Emissions in Iskandar Malaysia

This section presents GHG emission results for Iskandar Malaysia as simulated via the Extended Snapshot tool (ExSS). As mentioned earlier, to quantify GHG emissions in Iskandar Malaysia, a range of parameters (demography, economic growth, industry structure, energy, technology, transportation and land use) from Iskandar Malaysia's CDP, the 24 Iskandar Malaysia blueprints and other official documents are considered in the ExSS modelling. Three scenarios have been generated from the simulation:

1. 2005 – base year scenario
2. 2025 business as usual (BaU) scenario – target year with the development according to the CDP and existing development and environmental policies, without additional carbon mitigation measures
3. 2025 countermeasure (CM) scenario – target year with the development according to an assumed improvised CDP that adopts carbon mitigation options from the low carbon society blueprint

Based on the simulation result from the ExSS model, GHG emissions of Iskandar Malaysia in 2005 have been estimated to be 11.4 $MtCO_2eq$, and the value is projected to almost triple to 31.3 $MtCO_2eq$ in the 2025 BaU scenario (Fig. 7.8). With the introduction of the proposed 12 LCS actions (see Sects. 7.4.1 and 7.4.2) and their correspondent implementation programmes from the LCSBP-IM2025, increment of GHGs emission has been projected to slow down significantly, leading to an estimated emission level of 18.9 $MtCO_2eq$ for the 2025 CM scenario. As the industry sector is the key component in supporting the fast-growing region of IM, the sector will remain the highest emission sector contributing to between 35 and 53 % of the total GHG emissions in IM for all three base year (2005), 2025 BaU and 2025 CM scenarios.

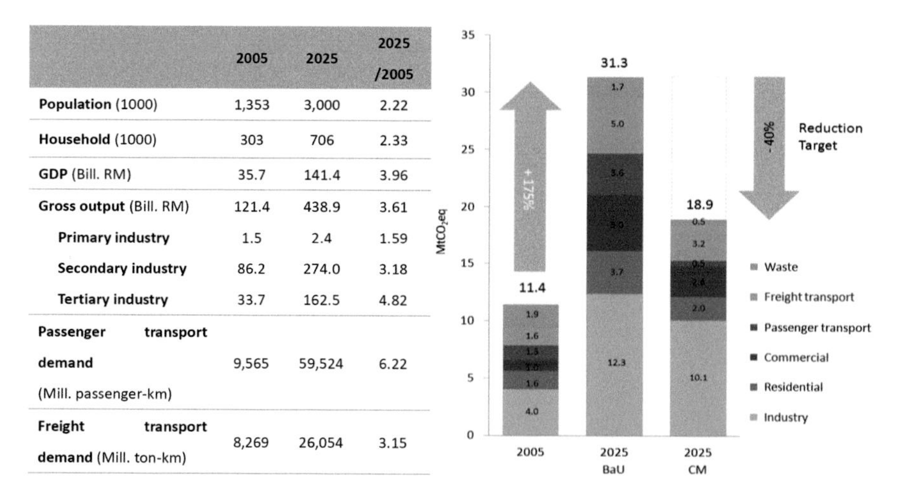

	2005	2025	2025 /2005
Population (1000)	1,353	3,000	2.22
Household (1000)	303	706	2.33
GDP (Bill. RM)	35.7	141.4	3.96
Gross output (Bill. RM)	121.4	438.9	3.61
Primary industry	1.5	2.4	1.59
Secondary industry	86.2	274.0	3.18
Tertiary industry	33.7	162.5	4.82
Passenger transport demand (Mill. passenger-km)	9,565	59,524	6.22
Freight transport demand (Mill. ton-km)	8,269	26,054	3.15

Fig. 7.8 Socio-economic scenario, GHG emission for base year (2005), business as usual (BaU) and countermeasure (CM) scenarios in 2025 (Source: UTM-Low Carbon Asia Research Centre 2013b, p. 1)

The result indicates that full implementation on the Blueprint's LCS programmes would potentially bring about a 58 % reduction of GHG emission intensity (over GDP) in 2025 compared to the 2005 level and a 40 % emission reduction in absolute terms from 2025 BaU (Fig. 7.8). This achievement is higher than the national commitment of 40 % voluntary carbon intensity reduction by 2020 and the Blueprint's initial target of 50 % reduction in intensity.

7.4.1 Structure of GHG Emission Mitigation Options

The LCSBP-IM2025 provides a sustainable green growth road map with 12 policy actions to move Iskandar Malaysia towards achieving its vision of a 'strong, sustainable metropolis of international standing' by 2025. The integration of two competing goals – 'strong' and 'sustainable' – in a single development vision poses great challenges to IM's growth policies and development planning. On the one hand, the urban region needs to develop a prosperous, resilient, robust and globally competitive *economy* (the 'strong' dimension); on the other (the 'sustainability' dimension), it needs to nurture a healthy and knowledgeable *society* that subscribes to low-carbon living and at the same time develop a total urban-regional *environment* that enables rapid economic growth but reduces growth's energy demand and carbon emission intensity. This calls for a holistic and integrated approach, involving policies and strategies on *Green Economy*, *Green Community* and *Green Environment* (Fig. 7.9), to decouple rapid growth from carbon emission in IM. Meeting this challenge has been the primary goal and underlying philosophy of the LCSBP-IM2025. Essentially, the Blueprint comprises two principal components:

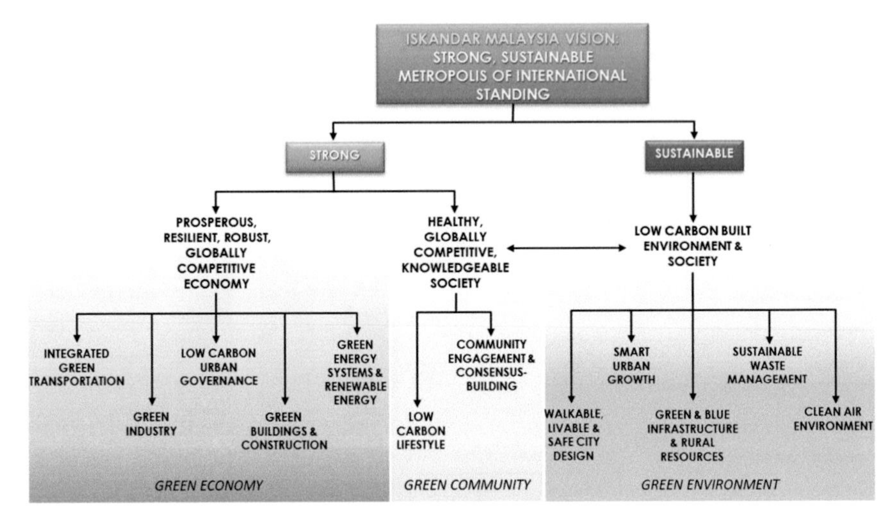

Fig. 7.9 Development of framework and scoping for LCSBP-IM2025 based on Iskandar Malaysia's development vision

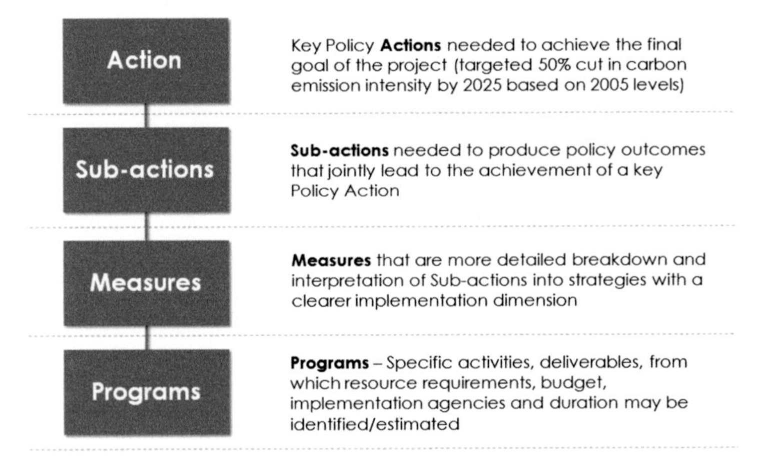

Fig. 7.10 'Work breakdown structure' (WBS) approach (Source: UTM-Low Carbon Asia Research Centre 2013a, p. 0–7)

1. Narrative on growth scenarios, policies, measures and programmes to achieve a minimum targeted 50 % reduction in carbon emission intensity by 2025 based on the 2005 level. Under this first component, 12 LCS Actions have been identified under the three themes: Green Economy, Green Community and Green Environment. To provide a clear framework for effective implementation of the policies, the 'work breakdown structure' (WBS) approach has been adopted where each of the 12 LCS Actions is collapsed into subactions and, in turn, into measures and detailed programmes (Fig. 7.10). A total of 281 LCS programmes have been identified for implementation in Iskandar Malaysia up to 2025 to meet

the GHG reduction target that has been set. For details of the 12 policy actions, subactions, measures and programmes (totalling some 400 policy items), readers are referred to the Full Report and Summary for Policymakers (SPM) of the Blueprint (UTM-Low Carbon Asia Research Centre 2013a, b).

2. Scenario-based modelling and projection of carbon emission reductions achievable using ExSS (see Sect. 7.3.2). With the implementation of the 281 LCS programmes identified in the Blueprint, a 58 % reduction in carbon emission intensity, from 319.33 tCO2eq/MYR1mil in 2005 to 133.66 tCO2eq/MYR1mil in 2025, has been projected for Iskandar Malaysia, which is higher than the 50 % reduction targeted at the outset.

7.4.2 GHG Emission Mitigation Options

As mentioned above, 12 mitigation options have been identified to lower carbon emissions of Iskandar Malaysia and transform the society of IM into a low carbon society. These have been organised under the three themes following the triple bottom line (TBL) of sustainable development, namely, Green Economy, Green Community and Green Environment. Collectively these 12 policy actions can potentially deliver a total of 12,758 ktCO_2eq direct emission reduction in 2025, accounting for a 40 % emission cut back from 2025 BaU. Table 7.1 shows the CO_2 reduction potential of each LCS Action. Three actions: *Action 5* (*Green Energy System and Renewable Energy*), *Action 6* (*Low-Carbon Lifestyle*) and *Action 1* (*Integrated Green Transportation*) jointly contribute to 57 % of total emission reduction; policymakers should pay more attention to these three actions and highlight them as higher-priority countermeasures that would help IM to cut carbon emission significantly.

7.5 Beyond Science and Policymaking: Implementing the LCSBP-IM2025

The LCSBP-IM2025 was prepared with its eventual implementation in mind from the outset. After its completion and launching at the UNFCCC's 18th Conference of the Parties (COP 18) in Doha, Qatar in November 2012 and its subsequent endorsement by the Prime Minister of Malaysia (who is also a Co-Chairman of IRDA) in December 2012 (Fig. 7.11), IRDA and the research team immediately looked into priority projects for implementation in Iskandar Malaysia for the 2013–2015 period. A series of intensive workshops were conducted between June and September 2013 and concluded in the formulation of the *Iskandar Malaysia Actions for a Low Carbon Future* (IRDA 2013), which outlines seven (7) specific Action-based projects plus three (3) special area-based projects (see Sect. 7.6.2) selected from the 281 programmes in the LCSBP-IM2025 for immediate implementation.

Table 7.1 Twelve mitigation options to lower the carbon emissions of Iskandar Malaysia

Mitigation options	Reduction[a] (ktCO2eq)	Percentage (%)
Green Economy	**6,937**	**54 %**
Action 1 Integrated Green Transportation	1,916	15 %
Action 2 Green Industry	1,094	9 %
Action 3 Low Carbon Urban Governance[b]	–	–
Action 4 Green Building and Construction	1,203	9 %
Action 5 Green Energy System and Renewable Energy	2,725	21 %
Green Community	**2,727**	**21 %**
Action 6 Low-Carbon Lifestyle	2,727	21 %
Action 7 Community Engagement and Consensus Building[b]	–	–
Green Environment	**3,094**	**25 %**
Action 8 Walkable, Safe and Livable City Design	263	2 %
Action 9 Smart Urban Growth	1,214	10 %
Action 10 Green and Blue Infrastructure and Rural Resources	392	3 %
Action 11 Sustainable Waste Management	1,224	10 %
Action 12 Clean Air Environment[c]	–	–
Total	**12,758**[c]	**100 %**

Source: UTM-Low Carbon Asia Research Centre 2013a, p. 0–5
[a]Contribution to GHG emission reduction from 2025 BaU to 2025 CM.
[b]Actions 3, 7 and 12 do not have direct emission reduction, but their effect is included in other Actions
[c]Since contribution of Action 10 includes carbon sink by forest conservation and urban tree planting, the total contribution of the 12 Actions is greater than difference of the GHG emissions between 2025 BaU and 2025 CM

Fig. 7.11 Launching of the LCSBP-IM 2025 at COP 18 (November 2012) and the Blueprint's endorsement by the Prime Minister of Malaysia (December 2012)

7.5.1 Selection of Priority Projects

The ten projects have been prioritised for implementation based on their institutional readiness (e.g. continuation of or extension to existing initiatives), relatively higher CO_2 reduction potential and lower implementation barriers, which include

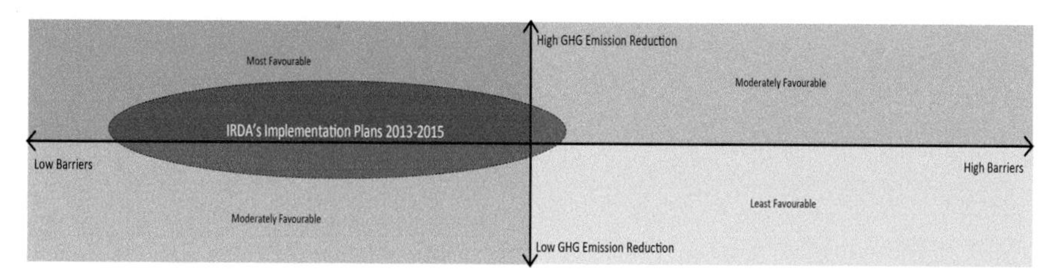

Fig. 7.12 Selection of priority LCS projects for implementation in 2013–2015 based on their relatively higher emission reduction potential and lower implementation barriers (Source: UTM-Low Carbon Asia Research Centre 2013c)

aspects of costs, human capital, institutional and legislative framework, society readiness (public acceptance), private sector buy-in and technology availability. Conceptually, these projects, when plotted in a four-quadrant plot along the axes of emission reduction potential and implementation barriers (Fig. 7.12), fall within the centre-upper-left region of the plot.

Another fundamental criterion underlying the selection of the ten implementation projects is that they should collectively cut across evenly all three main themes – Green Economy, Green Community and Green Environment – and the 12 LCS Actions of the Blueprint. To that end, a 'Project versus LCS Action mapping' exercise has been carried out, which shows a well-distributed coverage of all three main themes and ten out of 12 Actions of the Blueprint by the ten projects (Fig. 7.13). Successful implementation of these projects will be highly essential as positive demonstrations to the local and business communities in IM that will potentially boost their confidence, acceptance, ownership and support of the other LCS programmes in the Blueprint.

7.5.2 Selected Projects for Implementation in IM (2013–2015)

This section provides a summary of the ten projects that have been selected out of the 281 LCS programmes in the LCSBP-IM2025 and shows how actions supported by science can be, and are being, used to enable and realise reduction in carbon emissions in IM. For details of each project, readers are advised to consult the *Iskandar Malaysia Actions for a Low Carbon Future* booklet (IRDA 2013) from which the following project summaries have been extracted. The ten projects may be divided into seven specific LCS Action-based projects and three special area-based projects, as follows:

Seven specific Action-based projects:

1. Integrated Green Transportation – Mobility Management System
2. Green Economy Guidelines

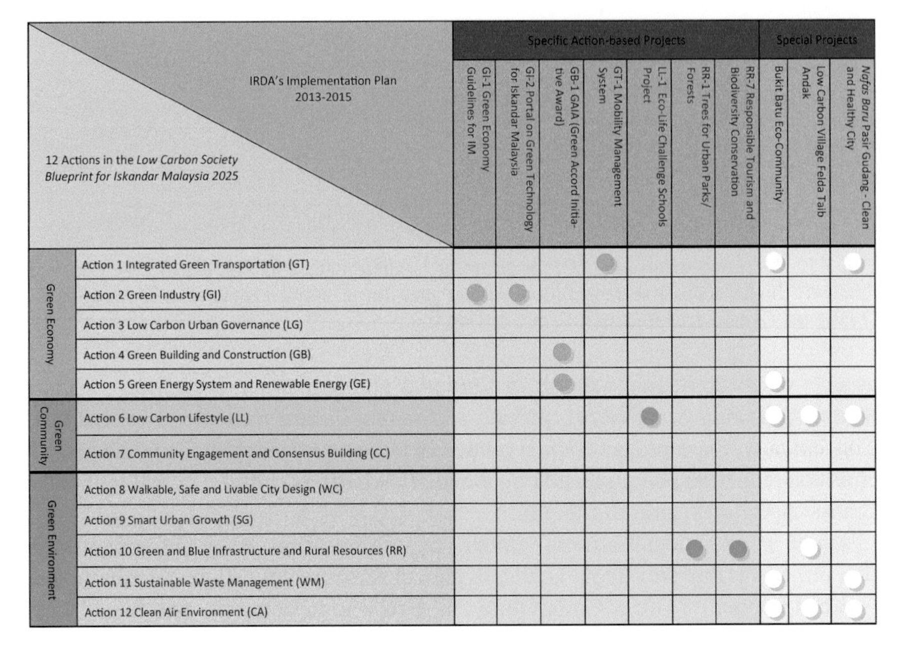

Theme	12 Actions in the *Low Carbon Society Blueprint for Iskandar Malaysia 2025* / IRDA's Implementation Plan 2013-2015	Specific Action-based Projects: GI-1 Green Economy Guidelines for IM	GI-2 Portal on Green Technology for Iskandar Malaysia	GB-1 GAIA (Green Accord Initiative Award)	GT-1 Mobility Management System	LL-1 Eco-Life Challenge Schools Project	RR-1 Trees for Urban Parks/ Forests	RR-7 Responsible Tourism and Biodiversity Conservation	Special Projects: Bukit Batu Eco-Community	Low Carbon Village Felda Taib Andak	*Nafas Baru* Pasir Gudang - Clean and Healthy City
Green Economy	Action 1 Integrated Green Transportation (GT)				●				○		○
	Action 2 Green Industry (GI)	●	●								
	Action 3 Low Carbon Urban Governance (LG)										
	Action 4 Green Building and Construction (GB)			●							
	Action 5 Green Energy System and Renewable Energy (GE)			●					○		
Green Community	Action 6 Low Carbon Lifestyle (LL)					●			○	○	○
	Action 7 Community Engagement and Consensus Building (CC)										
Green Environment	Action 8 Walkable, Safe and Livable City Design (WC)										
	Action 9 Smart Urban Growth (SG)										
	Action 10 Green and Blue Infrastructure and Rural Resources (RR)						●	●		○	
	Action 11 Sustainable Waste Management (WM)								○		○
	Action 12 Clean Air Environment (CA)								○	○	○

Fig. 7.13 Project versus LCS Action mapping exercise that shows even coverage across the Blueprint's three main themes and 12 Actions by the ten selected projects (Source: UTM-Low Carbon Asia Research Centre 2013c)

3. Eco-life Challenge Schools Project
4. Portal on Green Technology
5. Trees for Urban Parks
6. Responsible Tourism Development and Biodiversity Conservation
7. GAIA (Green Accord Initiative Award)

Three special area-based projects:

8. Bukit Batu Eco-Community
9. Low Carbon Eco Village FELDA Taib Andak
10. Special Feature: Smart City – Nafas Baru Pasir Gudang – Green and Healthy City

1. Integrated Green Transportation – Mobility Management System

The Iskandar Malaysia Mobility Management System (IMMMS) promotes sustainable transport and manages the demand for car use by changing travellers' attitude and behaviour. MMS coordinates information, services and activities to optimise the effectiveness of urban transportation. It is an innovative approach in managing and delivering coordinated and inclusive transportation services to customers, including the elderly, people with different abilities and low-income population. It is an online platform accessible through computers and smart phones, connecting citizens to the various modes of travelling within Iskandar Malaysia.

Project components include route and schedule information; trip/journey planner and travel optimisation; current travel conditions, alerts and avoidance; real-time transit arrival information; user travel analysis and system analysis.

2. Green Economy Guidelines

The Green Economy Guidelines look into areas of procurement, operations and supply chain management for businesses in order to minimise their impacts on the environment. The guidelines call for the government to look into the prospect of developing, adapting and revising current policies to support green growth through tax breaks, reducing perverse incentives and promoting and rewarding good practices for going green. Once the guideline is fully adopted by the various sectors within IM by the end of 2015, it will help to enhance the region's economic growth in tandem with environmental protection and conservation, supported by a green workforce and informed communities which generate positive impacts towards achieving IM's vision as well as contributing to GHG emission reduction in the region.

3. Eco-life Challenge Schools Project

The Children's Eco-life Challenge project (ELC) is an eco-household accounting project designed for students. The project is recommended as a supplement to the existing formal curriculum as a form of contextual learning, promoting systems-thinking and encouraging students to apply knowledge and skills learnt to real-life context. Through ELC, students monitor their own behaviour pattern as well as their families' in moving towards a low carbon lifestyle. Aspects included in ELC are energy consumption, waste generation and management, travel choice, frugal consumption and use of renewable resources. In 2013, the first batch of 22 Iskandar Malaysia-based UNESCO Associated Schools Project Network (ASPnet) primary schools participated in the inaugural ELC competition; the champion school was sent on an educational trip to Kyoto. The number of participating schools increased to 80 in 2014, and all 198 primary schools in IM (with a student population of over 184,000) are expected to be included under the ELC project in 2015. The complete Eco-life Challenge module and lesson plan are scheduled for a 2015 roll-out in IM. Updates on the implementation of the ELC project can be found at http://www.sustainableiskandar.com.my.

4. Portal on Green Technology

The Green Portal is a website and online platform where local communities, the government, private businesses, developers, investors and the wider public can access information related to green technology and the natural environment in IM. The portal is a one-stop virtual centre providing the latest news and information on green technology and LCS related topics, strategies, policies and guidelines. The portal also serves as a platform for promoting green employment and facilitating networking of a 'carbon literate' workforce to meet the growing needs of local, national and international green industries, notably those located within IM. The portal contains both historical and recent information on the natural environment such as policies on spatial/land use, shoreline planning, energy and waste management which will improve the knowledge of viewers and industry practitioners in both green technology and natural environment. The Green Portal is hosted in IRDA's current IM website at http://www.iskandarmalaysia.com.my/.

5. Trees for Urban Parks

'Trees for Urban Parks' aims to increase green lungs in urban areas in IM; enhance places for people to visit, exercise, recreate and socialise; and create habitats for attracting birds and small animals back into urban settings. The project calls for the retention as well as reintroduction of endemic tree species in urban parks and urban forests in IM. Key strategies to that effect include effective enforcement of the Trees Preservation Order (TPO) under the Town and Country Planning Act, 1976 (Act 172), promoting and encouraging planting of endemic trees among developers and monitoring and annual reporting of endemic tree planting. To begin with, IRDA has carried out a fairly comprehensive inventory of tree and plant species that are endemic to Johor and particularly to IM. Currently proposed tree planting programmes cover *Hutan Bandar MBJB* (MBJB Urban Forest) and *Taman Merdeka* in the Johor Bahru City limits.

6. Responsible Tourism Development and Biodiversity Conservation

Responsible tourism and biodiversity conservation inherently bring about simultaneous economic, social and ecological benefits and as such are candidate priority projects for advancing IM's LCS goal. Defining characteristics of responsible tourism are environmental education, community-led projects and promotion of resilient local economy. Building on the success of the first and second Eco-tourism Summits in 2012 and 2013, IRDA successfully led local communities and villagers in *Kampung Sungai Melayu* to actively champion the conservation of local natural resources (e.g. mangrove forests, grounds of migratory birds) and involve in sustaining and improving their economic livelihood. Starting as an IRDA-led project focusing on birding, taking advantage of the September–March migratory bird season, the annual project has taken its own momentum. Local communities are beginning to take stronger responsibility and pride in their natural environment, viewing it more than just a source of livelihood. IRDA now looks into transmitting the success formula to other coastal communities in IM.

7. GAIA (Green Accord Initiative Award)

The Green Accord Initiative Award or GAIA recognises and awards worthy companies and businesses in Iskandar Malaysia that have pioneered green and low carbon principles in their operation. While most companies comply with required environmental regulations or social requirements, some companies have initiated to look beyond regulatory compliance. This effort includes working with local communities in sectors of health and well-being, alleviating poverty, conserving the environment, and reducing carbon footprints. In this initial phase, GAIA will be looking specifically at green building development and companies that have adopted efficient energy system and implemented renewable energy approaches.

GAIA will be awarded to worthy development projects and buildings that have met local and international codes on green buildings, especially building design, and the application of green technology in their construction. GAIA is a soft incentive that will be tied to local and international rating tools such as Malaysia's Green Building Initiative (GBI), Japan's CASBEE and Singapore's Green Mark as well as other known assessment tools (e.g. LEED and BREEAM) to evaluate and recognise green buildings in Iskandar Malaysia. Work towards developing GAIA assessment criteria began in 2013 with the assessment tool being finalised in 2014. Evaluation begins in 2015 and will be conducted through collaborations with industry players and in consultation with all relevant parties in IM.

8. Bukit Batu Eco-Community

The Bukit Batu Eco-Community project aims to demonstrate how village communities whose current economic base predominantly revolves around oil palm and rubber plantations can achieve higher-value economic development within a low carbon society framework. The project seeks to improve villagers' lifestyle and financial status in and around the Kulai District in a sustainable manner via local employment, entrepreneurship and business co-ownership. The development emphasises the adoption of appropriate green technologies and various LCSBP-IM2025 recommendations to become a showcase model for other rural communities with rapidly urbanising contexts in IM. The initial phase of the proposed Eco-community of 1,214 hectares sits on a 4-ha site that is strategically located at the first exit of the North–South Expressway to IM, about 40 km northwest of Johor Bahru City. The first-phase development, funded via IRDA's Social Project Fund (SPF), comprises a business and marketing centre for local SMEs to market their products and services. Apart from yielding economic gains, the centre also provides various training, mindset change and social development programmes for the village communities. Environmentally, the centre will have the first rural green building that attempts to generate electricity from solar, wind and biomass sources; implement rainwater harvesting and promote 3R and green transportation.

9. Low Carbon Eco Village FELDA Taib Andak

Low Carbon Eco Village FELDA Taib Andak is a pioneer project that began in 2012 under Action 7 of the LCSBP-IM2025 to develop a model for low carbon community that incorporates the application of low carbon mitigation measures, such as energy-saving practices, use of oil palm biomass, 3R (reduce, reuse and recycle), production of green goods and reducing private transportation use. A key emphasis of the project is active community involvement in formulating and subsequently implementing a 12-action low carbon village blueprint towards promoting a low carbon lifestyle within the rural community. Successful programmes to date include organic waste composting, bamboo plantation, provision of recycling bins in each residential block, use of bicycles, social awareness programme and zero open burning. Continuous engagements with the community and regular monitoring of project implementation will be carried out by FELDA (Federal Land Development Authority), IRDA and Universiti Teknologi Malaysia.

10. Special Feature: Smart City – Nafas Baru Pasir Gudang – Green and Healthy City

Nafas Baru literally means 'new breath'. It is a programme mooted by IRDA together with *Majlis Perbandaran Pasir Gudang* (MPPG, Pasir Gudang Municipal Council) to rejuvenate Pasir Gudang to become a Green and Healthy City by 2025. Nafas Baru is in line with both the LCSBP-IM2025 and the *Iskandar Malaysia Smart City Framework*. The objective is to create 'smarter residents' in terms of resource planning and management through community actions where residents, the municipal council, industries and others work towards transforming Pasir Gudang into a clean, green, healthy and vibrant city.

With the aim of reducing carbon emission intensity by focusing on the three LCSBP-IM2025 pillars: Green Economy, Green Community and Green Environment, four (4) main initiatives have been identified to be implemented in the 2013–2015 period:

1. Green Industry

The green industry programme aims to make the existing industries in Pasir Gudang 'greener', low carbon and environmentally friendly without compromising their production and profit. As a start, ten industries of various types, sectors and sizes have been selected as pilot projects in which participating industries receive assistance to gauge their current status and readiness to adopt green industry initiatives. This will provide the basis for developing industry-specific low-carbon action plans.

2. Green Community Programme

This initiative aims to promote green community and green lifestyles among residents of Pasir Gudang through increasing the level of public awareness of climate change issues and LCS and encouraging them to live a lower-carbon lifestyle. Awareness campaign and teach-in programmes have already started, which focus on strategic aspects that are directly relevant to Pasir Gudang, including energy efficiency, 3R, composting, tree planting, smart travel choices, walking and cycling.

3. Integrated Solid Waste Management (Waste to Energy)

This programme seeks to enhance the implementation of the Integrated Solid Waste Management Blueprint 2009 in the Pasir Gudang area through developing an integrated and sustainable framework for managing solid wastes generated in the area. This is achieved through nurturing a participative and actively engaged public that is motivated to manage solid wastes in an environmentally and socially responsible manner; institutionalising a social and industrial solid waste management 'preferential framework' in the order of eliminate, reduce, reuse and use of advanced treatment and disposal technologies; and developing recycling and treatment technologies capable of generating beneficial by-products with zero or minimal emission.

4. Carbon Sequestration through the Trees Preservation Order (TPO)

This initiative aims to arrest and gradually reverse the decline in the carbon stock in trees in Pasir Gudang. To begin with, MPPG has identified 250 trees to be gazetted under the Trees Preservation Order (TPO) under the Town and Country Planning Act, 1976 (Act 172). The trees, covering 19 species with an age range of 8–30 years, have significant characters of being large and healthy, rare and unique and having substantial aesthetic, historic or tourism value. IRDA and MPPG are now working to price tag TPO gazette trees by identifying their carbon sequestration and monetary values, which will serve as a guide to future tree planting and urban landscape design in Pasir Gudang towards contributing to carbon emission reduction in Iskandar Malaysia.

7.6 Lessons Learnt

The Low Carbon Society Blueprint for Iskandar Malaysia 2025 is the first of its kind in Malaysia and, in the sense of its urban-regional scale application, perhaps among the few pioneering examples in Asia. The completion of the LCSBP-IM2025 and its launching at COP 18, Doha, in November 2012, are a major milestone in the *Project of Development of Low Carbon Society Scenarios for Asian Regions* that is sponsored by the Japan International Cooperation Agency (JICA) and Japan Science and Technology Agency (JST) under the Science and Technology Research Partnership for Sustainable Development (SATREPS) programme. The virtually no-time-gap selection and actual implementation of ten projects that collectively put ten of the Blueprint's 12 LCS Actions into *real action* in IM in the 2013–2015 period (see Sect. 7.5 above) are another significant achievement of the project. The project which officially commenced in July 2011 offers many valuable lessons especially in advancing scientific research on LCS into policymaking and, importantly, into actual implementation of the policies. A discussion of the key lessons learnt in the project is thus important and is in line with the objectives of the project under the SATREPS framework, which include the development of a methodology for creating LCS scenarios that are appropriate to Malaysia's context and dissemination of the methodology in the form of training programmes to other Asian countries. This section expands on the lessons identified in the project discussed in two earlier papers (see Ho et al. 2013a, b) and adds new revelations further gained since then.

The SATREPS funding framework necessitated at the outset of the project the setting up of a high-level Joint Coordinating Committee (JCC) that comprises top officials of key Malaysian Federal and Johor State government agencies that are relevant to the LCS project and their Japanese counterparts to oversee the implementation and review the progress and achievements of the project. It is found that this set-up indirectly boosts the government agencies' awareness of, and to some degree their commitment to, and buy-in of, the idea of LCS in urban and regional development. This may be important for the long-term advancement of the LCS agenda in the country, with strong endorsement from the central government, which potentially results in more effective GHG emission reduction.

In order to ensure LCS research that leads to effective LCS policies which are able to meet current policy needs, fulfil policymakers' expectations and fit into the wider policy framework, it is essential that a good inventory and understanding of existing policies across all government levels on economic and social development, environmental protection and climate mitigation are gained, in particular in terms of their interrelationship and of identifying policy gaps which the research should be designed to fill (see Fig. 7.4). Good understanding of the legal-institutional framework is also crucial to determine the form of LCS policies to be prepared, whether it should be a stand-alone policy or mainstreamed into existing policies (see UN-Habitat 2012), which influences the research process. In the case of the LCSBP-IM2025, a stand-alone LCS policy was prepared and subsequently mainstreamed into the local planning mechanism. In the final analysis, LCS research should be *policy oriented*, aiming at providing objective scientific evidences and concrete support for good LCS policies, which are in turn *research informed* and *evidence based*. Such integration and synergy potentially benefit both the research and policymaking sides, overcoming the situation of lack of communication between researchers and policymakers (UNCTAD Virtual Institute 2006) and building mutual trust between them, which opens up to more collaboration opportunities in the future for the creation of meaningful, implementable and effective LCS policies.

To also ensure that the LCSBP-IM2025 reflects as much as practicable the needs, concerns and aspirations of the entire IM communities, which potentially leads to higher level of awareness and ownership of the Blueprint and greater support for the implementation of the LCS programmes among the communities, it is learnt that continuous inclusive engagement of various stakeholders in IM through a series of focus group discussions (FGDs) is highly effective. FGDs have been designed into the research process at multiple stages where research findings and policy proposals were exhibited and actively discussed with stakeholders that range from Federal, State and local government agencies; industries; local businesses; civil societies, residents' associations and specific community groups; and various local NGOs and NPOs. Each FGD yielded useful feedback and opinions that were fed into the evolving policy proposals, which were fielded again in the subsequent FGD for further scrutiny by the stakeholders. Effectiveness of the FGDs is evidenced through the progressive improvement and refinement to each subsequent draft LCSBP-IM which began with 7 LCS Actions initially and expanded to 8, 10, 11 and the final 12 LCS Actions that provide the mainframe for the 281 LCS programmes to be implemented.

As the research progressed, an ever presence of 'science-policy gaps' was felt in terms of timescale (e.g. long-term versus short-term gains), priority (e.g. economic feasibility and budgetary concerns over social and ecological impacts) and practical considerations (e.g. institutional capacity and human capital to translate research into policy) between policymakers and researchers. While not all gaps were able to be patched as well as intended, it is learnt that having policymakers (committed IRDA officials) on-board the research team (see Fig. 7.5) helped significantly in identifying these issues as they cropped up and in promptly finding middle grounds. The inclusion of IRDA officials in the research team effectively brings the 'science/research realm' into the 'policy realm' and vice versa. This, to our knowledge, is rather uncommon; the more common research practice would be to periodically consult policymakers at several stages of the research process in which the policymakers' role tends to be advisory (reviewing and providing input, feedback, critique, etc.) instead of being continuously actively involved in shaping and conducting the research itself as in the case of the LCSBP-IM research.

It is further found that 'disciplinary gaps' exist even among academic researchers from different professional (e.g. social science versus pure science and engineering) and academic-cultural (e.g. Malaysian and Japanese research cultures and use of terminologies) backgrounds. An example of such a gap is the initial disagreement between planners who tended to take a more holistic and integrated view of policies and their interrelationship and engineers who tended to be precise about the boundaries and need for mutual exclusiveness between policies to avoid double counting in quantitative modelling of GHG emissions. Research team members need to be prepared to put in extra efforts and time to communicate and understand the other side's standpoint and smooth over any conflicts that arise. While disciplinary gaps are perhaps inevitable, it is found that working over them gives rise to perspectives and solutions that otherwise would not be thought of, thus leading to more creative and inclusive policymaking.

Effective communication of research evidences is vital; research evidences need to be communicated in straightforward languages, readable and graspable to policymakers who normally have very limited time. Furthermore, proposed policies need to 'appeal' to policymakers through, among others, identification of 'quick win' and 'low-lying fruits' programmes; emphasising social, health, air quality and environmental co-benefits of LCS programmes that will lead to potential public cost saving and greater public acceptance; outlining clearly direct implementation, resource allocation and benefit/cost implications; and showing sensitivity to institutional capacity and needs.

Towards ensuring high levels of buy-in from the government that result in speedy implementation of the LCS policies, it is learnt that strategic positioning and aligning of the policies in relation to the country's highest level, top priority policies are essential. In the case of the LCSBP-IM2025, the Blueprint emerged to be the first concrete policy that effectively and positively responds to the Prime Minister's COP 15 pledge to reduce the Malaysia's carbon emission intensity of GDP by 40 % by 2020 based on the 2005 emission level; the LCSBP-IM2025 demonstrates a potential reduction in emission intensity of GDP by 58 % in Iskandar Malaysia by 2025, based on the 2005 level. This translated into a high-profile endorsement of the Blueprint by the Prime Minister in December 2012, a month after the Blueprint's launch at COP 18. In addition, the LCSBP-IM2025 is also aligned to the Prime Minister's recent 'Science to Action' (S2A) initiative (see Sect. 7.3.1).

Another possible reason behind the almost immediate adoption and implementation of the LCSBP-IM2025 in IM in 2013 may be the status of the Co-Chairmen of IRDA, who are the country's Prime Minister and the Johor State's *Menteri Besar* (literally the Chief Minister). While this may be a given advantage in Iskandar Malaysia, in that IRDA's Co-Chairmen are statutory in nature under the IRDA Act, 2007 (ACT 664), the lesson learnt here is that it pays to have the top and most powerful politicians presiding over the area in which an LCS policy is to be implemented.

Apart from getting strong support from the highest level of the government, having strong leadership and committed officials at the local agency level who believe in the importance of scientific research in good policymaking is indispensable. Importantly for the LCSBP-IM2025, implementation agency level leadership and officials also consistently show deep commitment to advancing the LCS agenda in Iskandar Malaysia and are willing to engage with research institutions to see to it that good LCS policies are put in place. Through this, IM gets the benefits of having high-quality research backing of its LCS policy from Universiti Teknologi Malaysia, with strong expertise and technical support from Kyoto University, NIES and Okayama University.

Lastly, what worked well in Iskandar Malaysia may not necessarily work equally well in other urban regions. It is hoped that this sharing of lessons learnt in carrying LCS research through into policies and on-the-ground implementation, nonetheless, offers useful initial reference points for possible replication and/or adaptation in other countries or urban regions aiming to pursue a similar sustainable, low carbon growth path. No two urban regions or cities are the same; each will have to carve out its own model of LCS in relation to its specific economic, sociocultural, ecological and legal-institutional contexts. What is clearly evidenced by the successes of the LCSBP-IM2025 thus far is that developing countries, subject to adequate international funding and research and technological assistance from developed nations, and with good synergy between highly committed local research institutions and policymakers, are capable of crafting and putting in place implementable low carbon society policies that will eventually contribute to mitigating global climate change through real cuts in GHG emissions while still achieving a desired level of economic growth.

Acknowledgement This work is supported by JICA and JST under the Science and Technology Research Partnership for Sustainable Development (SATREPS) programme.

References

Ali G, Abbas S, Qamer FM (2013) How effectively low carbon society development models contribute to climate change mitigation and adaptation action plans in Asia. Renew Sust Energ Rev 26:632–638

de Oliveira JAP, Doll CNH, Suwa A (2013) Urban development with climate co-benefits: aligning climate, environmental and other development goals in cities. UNU-IAS, Yokohama

Gomi K, Shimada K, Matsuoka Y (2010) A low-carbon scenario creation method for a local-scale economy and its application in Kyoto City. Energ Policy 38:4783–4796

Ho CS, Ismail Ibrahim, Joeman BD, Muhammad Zaly Shah Muhammad Hussein, Chau LW, Matsuoka Y, Kurata G, Fujiwara T, Shimada K, Gomi K, Yoshimoto K, Simson JJ (2010) Low-Carbon City 2025: sustainable Iskandar Malaysia. Universiti Teknologi Malaysia, Johor Bahru

Ho CS, Matsuoka Y, Chau LW, Teh BT, Simson JJ, Gomi K (2013a) Blueprint for the development of low carbon society scenarios for Asian Regions – case study of Iskandar Malaysia. IOP Conf Ser Earth Environ Sci 16(1):012125

Ho CS, Matsuoka Y, Chau LW, Teh BT, Simson JJ, Gomi K (2013b) Bridging science and policy making: low carbon society blueprint for Iskandar Malaysia 2025. In: Proceedings of the 7th Southeast Asian Technical University Consortium (SEATUC) symposium. Bandung, Indonesia, 4–6 March 2013

IRDA (2013) Iskandar Malaysia actions for a low carbon future. Iskandar Regional Development Authority, Johor Bahru

IRDA (2014) Your guide to low carbon lifestyle in Iskandar Malaysia. Iskandar Regional Development Authority, Johor Bahru

JPBDSM (2012) Green neighbourhood planning guideline. Ministry of Urban Wellbeing, Housing and Local Government, Kuala Lumpur

KeTTHA (2009a) National green technology policy. Ministry of Energy, Green Technology and Water Malaysia, Putrajaya

KeTTHA (2009b) National renewable energy policy and action rlan. Ministry of Energy, Green Technology and Water Malaysia, Putrajaya

Khazanah Nasional (2006) Comprehensive development plan for South Johor economic region 2006–2025. Khazanah Nasional, Kuala Lumpur

Law of Malaysia, Iskandar Regional Development Authority Act, 2007 (Act 664). Percetakan Nasional Malaysia Berhad, Kuala Lumpur

Law of Malaysia, Town and Country Planning Act, 1976 (Act 172). Percetakan Nasional Malaysia Berhad, Kuala Lumpur

Li J (2011) Decoupling urban transport from GHG emissions in Indian cities – a critical review and perspectives. Energ Policy 39(6):3503–3514

MNRE (2009) National policy on climate change. Ministry of Natural Resources and Environment, Putrajaya

PEMANDU (2010) Economic transformation programme: a roadmap for Malaysia. Performance Management and Delivery Unit (PEMANDU), Prime Minister's Department, Putrajaya

New Partnership for Climate Resilient Development (2014) Retrieved from: www.science2action.my/index.php/item/136-new-pertnership-for-climate-resilient-development. 25 December 2014

Seto KC, Dhakal S, Bigio A, Blanco H, Delgado GC, Dewar D, Huang L, Inaba A, Kansal A, Lwasa S, McMahon JE, Müller DB, Murakami J, Nagendra H, Ramaswami A (2014) Human settlements, infrastructure and spatial planning. In: Edenhofer O, Pichs-Madruga R, Sokona Y, Farahani E, Kadner S, Seyboth K, Adler A, Baum I, Brunner S, Eickemeier E, Kriemann B, Savolainen J, Schlömer S, von Stechow C, Zwickel T, Minx JC (eds) Climate change 2014: mitigation of climate change. Contribution of working group III to the fifth assessment report of the intergovernmental panel on climate change. Cambridge University Press, New York

Skea J, Nishioka S (2008) Policies and practices for a low carbon society. Clim Pol 8:S5–S16

UN-Habitat (2012) Developing local climate change plans: a guide for cities in developing countries, cities and climate change initiative tool series. UN-Habitat, Nairobi

UNEP (2011) Decoupling natural resource use and environmental impacts from economic growth. A report of the working group on decoupling to the international resource panel. Available from: http://www.unep.org/resourcepanel/decoupling/files/pdf/Decoupling_Report_English.pdf

UNEP (2014) Decoupling 2: technologies, opportunities and policy options. A report of the working group on decoupling to the international resource panel. Available from: http://www.unep.org/resourcepanel/Portals/24102/PDFs/IRP_DECOUPLING_2_REPORT.pdf

UNCTAD Virtual Institute (2006) Research-based policy making: bridging the gap between researchers and policy makers. Joint UNCTAD-WTO-ITC Workshop on Trade Policy Analysis, Geneva, 11–15 Sept 2006. Available from: http://vi.unctad.org/tda/papers/tradedata/tdarecs.PDF

UTM-Low Carbon Asia Research Centre (2013a) Low carbon society blueprint for Iskandar Malaysia 2025 – summary for policymakers, 2nd edn. UTM-Low Carbon Asia Research Centre, Johor Bahru. Available from: http://www.utm.my/partners/satreps-lcs/publications/

UTM-Low Carbon Asia Research Centre (2013b) Low carbon society blueprint for Iskandar Malaysia 2025 – Full report. UTM-Low Carbon Asia Research Centre, Johor Bahru. Available from: http://www.utm.my/partners/satreps-lcs/publications/

UTM-Low Carbon Asia Research Centre (2013c) A roadmap towards low carbon Iskandar Malaysia 2025. UTM-Low Carbon Asia Research Centre, Johor Bahru. Available from: http://www.utm.my/partners/satreps-lcs/publications/

World Bank (2010) Cities and climate change: an urgent agenda. World Bank, Washington

Part III
Best Practices and Recommendations in Each Sector to Make It Happen

Chapter 8
Low-Carbon Transport in India

Assessment of Best Practice Case Studies

P.R. Shukla and Minal Pathak

Abstract India is the world's fourth largest emitter of greenhouse gases. Transport contributes 13 % of India's GHG emissions (MoEF. India: green house gas emissions 2007, Indian Network for Climate Change Assessment (INCCA), Ministry of Environment and Forests (MoEF). Government of India, New Delhi. Accessed 13 Sept 2013, 2010). Driven by rising population, income, and urbanization, under a business-as-usual scenario, India's energy demand from transport is projected to increase sixfold in 2050 from current levels. This has vital impact on key national sustainable development indicators like energy security and air pollution. In response, several national and subnational policies and measures were initiated to ameliorate the adverse impacts of transport decisions on sustainability. These include national policies and programs for fuel efficiency, low-carbon technologies, investments in public transport infrastructure, and climate change mitigation. These aside, several bottom-up interventions that are initiated locally are showing promise.

This chapter offers an overview of transport sector in India and presents selected best practice case studies that identify good practices. Evidently, the challenge is to replicate and scale up these practices to gain sizable CO_2 mitigation together with co-benefits vis-à-vis various national sustainable development goals. The assessments show that successful implementation of national policies at the subnational level requires widely agreed goals and targets and support from the national government. The support can be in the form of capacity building, technology, or finance. In the overall, the chapter argues for (1) integrating transport policies with local, national, and global objectives, (2) a comprehensive assessment of the impacts (co-benefits and risks) of policies and project from the planning to the post-implementation stage, and (3) cooperation and knowledge sharing among cities and regions facilitated by the national government for cross-learning and transfer of best practices. The lessons from these studies provide important

P.R. Shukla
Indian Institute of Management, Ahmedabad, India

M. Pathak (✉)
CEPT University, Ahmedabad, India
e-mail: minal.pathak@cept.ac.in

S. Nishioka (ed.), *Enabling Asia to Stabilise the Climate*,
DOI 10.1007/978-981-287-826-7_8

learnings for designing policies and projects elsewhere including other developing countries.

Keywords Low carbon • Best practice • Transport • Co-benefits • Replicability • Policy

Key Message to Policy Makers

- Transport "best practices" deliver sustained carbon mitigation and co-benefits.
- Best practice cases include policies and projects which are replicable and scalable.
- Case studies demonstrate contextual effectiveness of methods and practices.
- Best practices align sustainability and low-carbon goals spatially and temporally.
- Rational transport system needs integration of inter- and intracity transport choices.
- Technology cooperation and carbon finance are vital for replication and scalability.

8.1 Introduction

This chapter examines selected case studies which represent "best practice" vis-à-vis sustainable low-carbon transport policies, measures, technologies, and investments in India. The term "best practice" is used contextually. The central idea for this chapter is not to look for a perfect blueprint of a policy or plan but to critically evaluate what has been or is being implemented and is proven to work. The key "best practice" criteria considered in assessing the case studies are clear vision, assessment of delivered or demonstrated co-benefits of low-carbon choices, as well as their replicability and scalability. The case studies are selected, keeping in view the diversity of transport sector and the multiple interfaces of the sector with development and environment.

The overarching transport policies are framed, mandated, and facilitated by the national government. The implementation in most cases involves subnational governments or sector-specific departments or organizations. The projects are implemented by agencies, including those operated under public-private partnership (PPP). The case studies are therefore selected belonging to both policy and project domains. Successful low-carbon integration between national and subnational governments is essential for successful implementation of policies which requires integration between national policies and subnational initiatives

(Matsumoto et al. 2014). The cases selected here look at both national and subnational initiatives on low-carbon transport policies and infrastructure.

The three case studies on policies include major initiatives by the Government of India to improve local environment and/or energy security, but also have bearing on greenhouse gas emissions from transport. The three project case studies include new areas where either the first project is recently initiated (e.g., dedicated freight corridor), the first project is identified (e.g., high-speed rail), or an initial set of projects are already under implementation (e.g., bus rapid transit).

8.1.1 Current Transport Scenario in India

India's transport sector is a rapidly growing sector and contributes 6.4 % to the GDP of the country. The sector is largely oil dependent and accounts for 13 % of the country's energy-related CO_2 emissions (MoEF 2010). Crude oil imports have been increasing steadily and making India the third largest oil importer globally. Nearly 80 % of India's current crude oil consumption comes from imports raising challenges of national energy security.

Intercity transport is mainly met by road (88 %), rail (11 %), and a limited share of air transport. Indian railways are among the largest rail networks globally and transport 23 million passengers and 3 million tonnes of freight daily (GoI 2015). Despite its extensive network, railways are faced with issues of capacity constraints and poor infrastructure. The share of rail has dropped from over 40 % in 1970 to 11 % in 2010 due to high competition from road transport. Similarly, rail dominated freight transport in India; however, this share is on the decline in recent years.

In urban areas, road transport dominates. Present status of urban transport is characterized by increasing trip distances, increasing share of private motorized transport, and declining share of public and non-motorized transport. These trends are leading to increasing problems of poor air quality, road safety, noise, and congestion.

8.1.2 Transport Scenarios for India

India is witnessing a unique period of population growth, economic growth, and urbanization. A third of India's population lives in urban areas. Urban population is expected to grow in the future, and by 2050, half of India's population is projected to reside in cities (UN 2014). India's GDP is also expected to grow at a healthy rate with per capita incomes reaching USD 15,842 (2010 prices) in 2050. Population, income, and urbanization are expected to drive vehicle ownership, travel demand, and freight transport demand.

Intercity travel demand will increase by 4.3 times between 2010 and 2050. In business-as-usual (BAU), this demand will be met by road-based transport and a growing share of air transport resulting in a higher energy demand resulting in challenges of national energy security and greenhouse gas emissions. In cities, increasing travel demand, reliance on private motorized modes, and declining share of public transport and non-motorized modes will increase energy demand and GHG emissions from cities (Dhar et al. 2013).

Under a BAU scenario, oil will dominate as the energy source, despite a minor diversification into natural gas, electricity, and biofuels. Increasing electrification of intercity rail, urban rail, and freight transport will increase electricity demand from transport. Transport emissions in the BAU are expected to reach around 1 billion tCO2 in 2050—an increase of 5.5 times increase from 2010 levels (Dhar and Shukla 2014). It is increasingly becoming clear that the BAU will not deliver the desired level of GHG mitigation. For policy makers in the Indian transport sector, this growth poses multiple challenges. Besides the impact on climate change, this raises other issues on how to offer wider mobility access at affordable rates, limit the health impacts of air pollution, and reduce traffic congestion and dependence on fossil fuels.

8.1.3 Need for Assessment

Concerns in developing countries exist regarding the costs imposed by mitigation targets and their impact on economic growth (Olsen 2013). The "co-benefits approach" helps identify actions that balance the short-term development concerns with long-term goals of climate change mitigation (IGES 2011; Creutzig and He 2009). Opportunities exist to mitigate GHG emissions from India's transport sector and facilitate sustainable mobility by integrating transportation policies with environment, development, and climate change policies. Key interventions include reducing travel demand through planning and sustainability measures, a shift of passenger and freight transport from road-based modes to rail and from private transport to public transport and non-motorized transport in cities, and increase penetration of alternate fuels and vehicles including electric vehicles and hybrid vehicles. These measures will also diversify the fuel mix with a higher share of electricity, natural gas, and biofuel (Dhar et al. 2013).

The sustainability focus is evident in policies of the Government of India. For instance, India's National Action Plan on Climate Change (NAPCC) highlights a mix of measures, including higher share of public transport, penetration of biofuels, and significant improvements in vehicle efficiency (GoI 2008). Several cities are proactively initiating infrastructure investments in mass transit, urban planning for better land use transport integration, and upgrading existing public transport. These policies and interventions have reduced GHG emissions and at the same time have delivered social and environmental benefits. Since these are limited to few cities, they have not realized the desired mitigation potential. Evidently, there is scope for replication to deliver higher emission reduction and deliver wide-ranging economic, social, and environmental benefits.

At the same time, some of these initiatives are beset with challenges during planning and implementation. It is essential to carry out a comprehensive assessment of good practices for three reasons: (1) this assessment can help highlight the mitigation potential and other benefits to guide policy makers in replication or scaling up, (2) it can highlight unique approaches or co-benefits, and (3) it can help understand challenges during planning and implementation which can be integrated during the next stage to avoid adverse impacts post-implementation.

For instance, the successful implementation of the Auto Fuel Policy 2003 catalyzed the development of a roadmap for further improvement till 2025. The success of transport initiatives in cities can facilitate cross-learning among subnational governments and help to bring in measures early. As an example, successful implementation of a mass transit system in a city can deliver useful lessons to subnational governments on developing mobility plans and leveraging finance for implementation. It is essential therefore to take critical and comprehensive assessment of objectives and impacts to guide future policies to better align these with development goals.

The central idea of this chapter is to look at selected case studies and highlight the success factors and critically examine issues in order to make informed decisions for replication in future. The paper is divided into four sections. After the introduction section, the second section outlines the key transport policies and plans in India. These include existing and proposed policies, planned investments including major infrastructure projects, and urban initiatives. A detailed assessment of case studies is described in Sect. 8.3. The final section concludes with the key highlights from the case study assessment.

8.2 Transport Policies in India

The Government of India has initiated several policies and initiatives for the transportation sector with the objective of enhancing passenger mobility, improving logistics of freight transport, increasing rail use by improving efficiency, raising the average speed, promoting low-carbon transport, and at the same time improving energy security and local benefits of air quality and congestion (Table 8.1). Cities have initiated urban transport initiatives including infrastructure for public transport and non-motorized transport and urban planning and zoning interventions to facilitate transit-oriented development.

Transport sector takes up a share of 45 % in the total infrastructure investments in India. There are plans to increase investments from 2.6 % of GDP between 2006 and 2011 to 3.6 % of total GDP in the period between 2018 and 2022. The Government of India policies highlight rapid expansion and modernization of transport infrastructure. Some of these include expansion and upgradation of roads and highways, reducing congestion in railways, electrification of rail corridors, investments in dedicated freight corridors, and expansion of air infrastructure

Table 8.1 Overview of selected transport policies in India

Sector	Policy/plan	Highlights
Urban transport	National Urban Transport Policy	Enhancing mobility to support economic growth and development
		Reduce environmental impacts
		Enhancing regulatory and enforcement mechanisms
	National Mission on Sustainable Habitat	Submission under India's National Plan on Climate Change
		One of the key components is promotion of urban public transport
Alternate fuels and vehicles	National Policy on Biofuels	5 % blending of ethanol in petrol in 20 states and eight union territories
		Financial incentives
		Waiver on excise duty for bio-ethanol and excise duty concessions for biodiesel
	National Electric Mobility Mission Plan	Investments in R&D, power, and electric vehicle infrastructure
		Savings from the decrease in liquid fossil fuel consumption
		Substantial lowering of vehicular emissions and decrease in carbon dioxide emissions by 1.3–1.5 % in 2020
		Phase-wise strategy for research and development, demand and supply incentives, manufacturing and infrastructure upgrade
Intercity passenger transport	High Speed Rail Project	High Speed Rail Corporation of India Limited (HSRC) formed for development and implementation of high-speed rail projects
		2000 km high-speed railways network (HSR) by 2020
		14 corridors identified
Efficiency	Fuel Economy Standards for cars	Binding fuel economy standards starting 2017
		Fuel efficiency improvement in cars by 10 % in 2017
		20 % in 2022
	Auto Fuel Policy	30 new cities are planned to move to Euro IV by 2015
		Euro V in the entire country by 2020
Freight	Dedicated freight corridors	Double employment potential in 5 years (14.87 % CAGR)
		Triple industrial output in 5 years (24.57 % CAGR)
		Quadruple exports from the region in 5 years (31.95 % CAGR)

investments in high-speed rail and mass transit in cities. Improving water-based transport is now receiving some attention, and this has been mentioned as one of the focus areas in the National Urban Transport Policy.

Table 8.2 Chronology of transport initiatives implemented

Year—measure implemented
1991—First set of mass emission norms for all vehicles introduced
1995—Catalytic converters made compulsory
1995—Unleaded petrol introduced in Delhi
1996—Diesel with 0.5 % S introduced in four metros and Taj Trapezium
1997—Low-sulfur diesel (0.25 %) in Delhi and Taj Trapezium
1998—Low-sulfur diesel (0.25 %) in three metros
1999—Euro I equivalent norms for passenger cars in Delhi
2000—Auto Fuel Policy Committee formed; unleaded petrol in the country; low-sulfur diesel
(0.25 %) in the country; (0.05 %) in four metros
2000–2001—Euro II equivalent norms for passenger cars in four metros
2002—All public transport converted to CNG in Delhi
2003—Phase out of old taxis
Three-wheelers to CNG in Mumbai
2005—Low-sulfur diesel (0.05 %) in the entire country; (0.035 %) in metros
2005—Euro III equivalent norms for all cars in seven megacities
2008—BRTS becomes operational in Delhi
2009—BRTS becomes operational in two other cities
2010—Low-sulfur diesel (0.035 %) in the entire country; (0.0005 %) in ten metros
2010—Euro IV equivalent norms in major cities; Euro III equivalent for the rest of the country
2011—Delhi Metro Phase II completed
2012—National Electric Mobility Mission Plan announced
2013—Ahmedabad BRTS ridership reaches record high
2014—Dedicated bicycle track in Diu
2014—Low-carbon comprehensive mobility plan for three cities
2015—Electric rickshaws legalized in Delhi

Compiled from: GoI (2003), CPCB (2008)

Emerging policies highlight the focus on multiple benefits of meeting the transport demand and delivering environment and development benefits (Table 8.1). An example is the recent initiative to develop high-speed rail corridors in the country (GoI 2014a) which is expected to benefit cities along major corridors by improving their connectivity.

Historically, transport interventions in India have been driven by various push factors. For instance, in Delhi, a public interest litigation regarding air pollution prompted a Supreme Court directive authorizing the conversion of public transport to CNG. This was a landmark achievement as Delhi's success prompted several other cities to bring in CNG vehicles. Similarly, the success of electric auto-rickshaws in Delhi was driven by favorable economics and not necessarily government intervention (Shukla et al. 2014). Table 8.2 documents the range of policies and interventions on improving air quality that were been implemented successfully since the 1990s.

8.3 Transport Policy at the National and Subnational Levels

8.3.1 Selection and Assessment Criteria

Assessment of best practice is a popular approach, and literature has focused on different facets of best practice research. The terms "best practice" and "good practice" are also debated. The central idea for this chapter is not to look for a perfect blueprint of a policy or plan but to critically evaluate the policies implemented or under implementation that are showing initial benefits. Vesely (2011) classifies good practices into (1) those that depend on functionality that were successful and generated replicable outcomes, (2) practices that emphasized a unique methodology that helped to achieve the objectives, and (3) practices where new approaches were introduced.

For the selection of case studies, we followed a stepwise method. The first step included listing potential case studies covering major transportation subsectors. These included policies, programs, and projects at the national, regional, and city levels. We used a broad-brush method to evaluate these on the three criteria based on the information available: (1) clear vision, (2) evidence or potential of reducing GHG emissions, and (3) delivered or demonstrated the potential of economic development and/or environmental advantages. The idea was to consider diverse case studies from national and subnational levels and from different subsectors of the transport sector—passenger, freight, technology, infrastructures, as well as policies.

Few peer-reviewed studies that comprehensively examine these policies are available. The assessment in case studies therefore relies on peer-reviewed studies (where available), research, gray literature including project reports, reports from think tanks, or other organizations that analyze experiences, published case studies, as well as official documents. Six case studies were studied for impacts at various dimensions and ongoing or post-implementation issues. For developing criteria, we referred to other similar assessments available (Vesely 2011; GGBP 2014). The cases were then assessed for (1) vision and impact (whether the policy/project had clear objectives and was implemented successfully), (2) replicability (whether the practices were scaled up or replicated in other contexts or have the potential for replication), and (3) co-benefits (the intervention has delivered or has the potential to deliver co-benefits). The range of benefits includes GHG mitigation, local air quality benefits, and social and development impacts. The first three case studies are implemented, and we have attempted to draw insights on impacts and challenges. The next three case studies are emerging practices and have the potential to deliver low-carbon sustainability benefits. We believe the insights coming out of each case will enhance understanding of these interventions to enable replication and deliver the ultimate goal of a low-carbon transport transition for India.

8.3.2 *Case 1: Delhi Metro*

The Delhi Metro Rail Corporation Limited (DMRC) was established to implement the construction of a mass rapid transit system in Delhi. The objective was to develop a mass transit system to enhance mobility and simultaneously to ease congestion and reduce air pollution in Delhi.

The first phase of the metro corridor with a length of 65 km was completed in less than 3 years. An additional 125 km in Phase 2 became fully operational in 2011, taking the present network to 193 km covering 140 stations. The infrastructure covering four phases totaling 245 km is expected to complete by 2021. The project was funded with a joint contribution of Japan International Cooperation Agency (JICA), joint equity contribution by the national and state governments, and a small proportion coming from property development (DMRC 2015). The Delhi Metro has not only improved connectivity within the city but has also improved transport integration through its airport-city link and regional connectivity through its planned connections to towns in the neighboring state of Haryana.

8.3.2.1 Impacts

Delhi Metro has a daily ridership of 2.6 million passengers (DMRC 2014b). A recent study has reported that about 0.3 million vehicles have been taken off the road due to the introduction of the Delhi Metro (CRRI 2011). Expansion of the metro network delivered air quality benefits of reduced NO_2, CO, and $PM_{2.5}$ (Goel and Gupta 2014). In 2011, shifting of commuters from road-based transportation to metro rail in Delhi saved 1320 tons of NOx, 107 tons of particulate matter, and over 3880 tons of CO_2 (Sharma et al. 2014).

This is the first urban rail CDM project globally and has achieved significant reductions in GHG emissions. This is also a landmark project for the country as it has already registered three successful projects under the Clean Development Mechanism (CDM) (DMRC 2014a). These include the carbon credits from regenerative braking, the Modal Shift Project, and the Energy Efficiency Project under CDM and Gold Standard which are expected to reduce approximately 570,000 tCO_2 annually. The project saved 90,000 tons of CO_2 from regenerative braking between 2004 and 2007 and continues to claim credit. Increasing ridership, modal shift, and energy conservation practices will deliver further mitigation benefits in the future (Sharma et al. 2014).

In response to the success of DMRC, the Government of India has submitted the MRTS Program of Activities (PoA) to The United Nations Framework Convention on Climate Change (UNFCCC). The PoA will cover a series of rail-based MRTS projects (like metro rail, LRT, monorail) implemented across India. The objective of MRTS PoA by DMRC is to promote implementation of mass transit systems to reduce GHG emissions and support with implementation for the construction of an MRTS projects by providing fast-track carbon funding and risk-free registration of future projects (UNFCCC 2014).

Khanna et al. (2011) carried out scenario analysis to demonstrate that a rail-dominated mass transit system in Delhi can deliver 61 % reduction in energy use compared to 31 % reduction for a bus-dominated system. A cost-benefit analysis of the Delhi Metro calculated a 22 % social rate of return, a financial rate of return of 17 %, and an economic rate of return of 23.9 % including gains from air pollution reduction (Murty et al. 2006). The study reported that Delhi Metro generated benefits to the stakeholders including citizens and government; however, other transport providers suffered from income losses (ibid).

The implementation of the Delhi Metro has resulted in social impacts including relocation of people and reduced accessibility of the relocated low-income households (Tiwari 2011). The DMRC faces challenges due to land acquisition issues. Metro infrastructure projects are being planned or are under construction in nearly 20 cities—several of which will follow the Delhi model. It is crucial to address issues of equity and development to minimize adverse social impacts during project implementation. An additional concern is the vulnerability of cities and infrastructures to the risks from climate variability, especially extreme weather events (Garg et al. 2013; Pathak et al. 2014). These considerations should be factored in as far as possible into planning of long-term transport infrastructure.

8.3.2.2 Replication and Scalability

The key lessons emerging from the Delhi Metro case study are as follows: (1) - low-carbon mobility projects can leverage financing through the international carbon market, (2) large infrastructure projects provide a good opportunity for technology cooperation between a developed and a developing country, (3) public transport projects, if implemented effectively, can deliver mitigation benefits and lead to environmental co-benefits from reduced congestion and improved air quality, and (4) large infrastructure projects should make all efforts to minimize adverse social impacts during planning and implementation.

8.3.3 Case 2: Auto Fuel Policy (AFP)

The Government of India set up the Auto Fuel Policy Committee in 2000 to prepare a policy for setting up of emission norms and fuel quality standards in the country and to provide a roadmap for its implementation (GoI 2003). In addition, the policy recommended improvement of fuel economy, reducing pollution from in-use vehicles, submitting vehicle for inspection and maintenance, and augmenting public transport.

The Committee recommended a progressive implementation of fuel quality norms and Euro equivalent exhaust emission norms for vehicles. Taking into account technical, financial, and institutional considerations, the roadmap suggested the implementation in a phased manner in the country starting from the

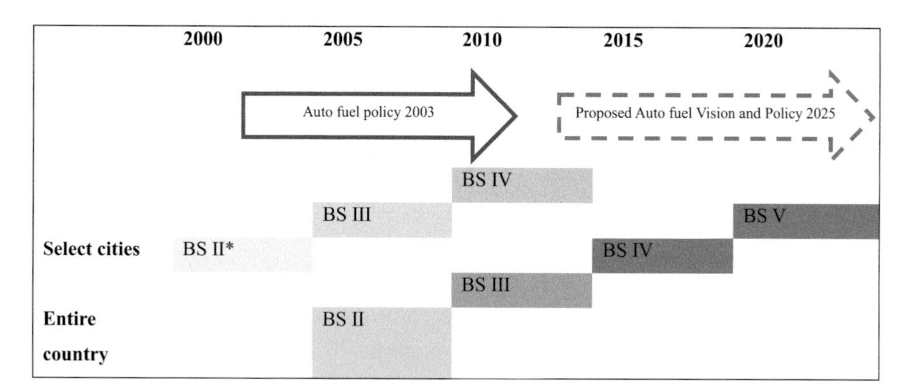

Fig. 8.1 Vehicular Emission norms in India: Implementation and future roadmap of India's Auto Fuel Policy (Sources: Authors. Adapted from: GoI 2003; 2014c; ICCT 2013a, b. Note: BS refers to Bharat Stage emission norms. These norms are broadly equivalent to Euro norms. This roadmap is for four wheelers including passenger cars, light commercial vehicles and heavy duty diesel vehicles)

implementation of Bharat Stage II fuel/emission norms in 2005 and progressively going to Euro IV in 2010. The implementation was to be initiated in 11 major cities followed by the rest of the country. Subsequently, in 2014, the Government of India outlined the Auto Fuel Vision and Policy that laid down future vehicle and fuel improvement roadmap for the country till 2025 (GoI 2014b). According to the roadmap, Bharat Stage (BS) IV will be adopted by 63 cities by the end of 2015 and adopted countrywide by April 2017. The implementation roadmap is shown in Fig. 8.1.

8.3.3.1 Impacts

Following the recommendation of the Auto Fuel Policy (GoI, 2003), implementation emission and fuel quality standards have been progressively modified in India. Euro I equivalent norms were introduced in the year 2000 starting with 11 identified cities including metros and subsequently implemented across the country. Subsequently, Euro II and Euro III equivalent norms were also introduced across the country in the year 2003 and 2005. India's Auto Fuel Policy report brought about significant air quality benefits compared to the "no-policy" scenario.

A detailed assessment showed that actions outlined in the policy were largely implemented in accordance with the policy roadmap (ICCT 2013a). There has been a reduction in sulfur content in gasoline (from 2000 ppm to 150 ppm) and diesel (from 10,000 ppm to 350 ppm). In selected 20 cities, sulfur content has been brought down to 50 ppm. As a result, SO_2 levels have come down significantly in Indian cities and remain well within WHO standards (CPCB 2012). Other improvements include a reduction in benzene levels in gasoline and increase in the use of zero-sulfur CNG and LPG in buses and auto-rickshaws. A recent study reported a significant decrease in PM_{10} vehicular emissions throughout the previous decade and a slowing of growth of NOx emissions. This resulted in 21,500 deaths avoided from reduced $PM_{2.5}$ emissions (ICCT 2013a). The clear vision, process of implementation, and success of the AFP 2003 helped replicate and upscale Auto Fuel Policy Vision 2025.

8.3.3.2 Replication and Scalability

Despite the progressive vision and roadmap and successful implementation of the policy, India's fuel quality and emission norms still lag behind international standards, and air quality remains a serious problem in all parts of the country. Efforts are required to advance the implementation of BS IV and V than scheduled in the Policy Vision of 2025. A study has shown that accelerating implementation of stringent fuel quality and vehicle emission standards in India will avoid approximately 48,500 premature deaths. This would impose an additional cost of USD 45 billion; however, the economic benefits of these avoided deaths would amount to USD 90 billion, making a strong case for advancing the roadmap (ICCT 2013b).

When the Auto Fuel Policy Committee was formed in 2000, the focus was to address air quality issues in the country. However, a decade later, energy security and climate change are simultaneous national priority issues. Recently, the government has proposed fuel economy standards for cars starting 2017 (BEE 2014). According to this draft proposal, fuel economy of cars will improve by 10 % in 2017 and 15 % by 2020. The Auto Fuel Policy Roadmap 2025 can consider conjoint mitigation of local pollutants and GHG emissions. The key lessons from the case study are as follows: (1) clear vision and targets at the national level were important factors in successful implementation; (2) a phase-wise implementation roadmap taking into account existing technical, institutional, and financial considerations; (3) consideration of stakeholders' concerns; and (4) future policies can aim for conjoint mitigation of local air pollutants and greenhouse gas emissions.

8.3.4 Bus Rapid Transit System

The National Urban Transport Policy was announced by the Ministry of Urban Development (MoUD) in 2006. Key focus areas of the policy included encouraging use of public transport and facilitating the introduction of high-quality multimodal public transport systems (MoUD 2006). Simultaneously, the National Urban Renewal Mission (NURM) committed substantial funds for bringing about improvements in urban infrastructure which included funds for BRTS projects in the country. Presently, over a dozen Indian cities have BRTS systems which are operational, and few other cities are planning to construct BRTS. The BRTS in Ahmedabad is among the most successful systems in the country. The BRTS in Ahmedabad was introduced in 2009 in order to promote public transport, to reduce the rapid growth of private vehicles, and to ease congestion. Based on a PPP model, the BRTS was implemented in a phased manner.

8.3.4.1 Impacts

The Ahmedabad Bus Rapid Transit System (BRTS) network has grown rapidly over the years and now operates a 52.5 km network with another 36 km under planning. The ridership started from 18,000 trips per day and has currently reached 0.1 million trips per day. In addition to increased ridership and revenue, the BRTS has resulted in other co-benefits including reduced travel time and other environmental benefits (NIUA 2014). Ahmedabad BRTS has improved mobility choices and reduced travel time (Rogat et al. 2015). The system has led to a reduction in CO_2 emissions and at the same time delivering social and economic benefits through enhanced mobility (UNFCCC 2015). The success of Ahmedabad BRTS encouraged several cities to implement BRTS based on the Ahmedabad model.

8.3.4.2 Replication and Scalability

A modal shift from private to public transport at lower cost can be facilitated by making public transport more attractive with a comprehensive plan that facilitates intermodal integration (Bubeck et al. 2014). The construction of two metro corridors in the city is commenced, the first phase of which is expected to initiate operations in 2018. Efforts are under way to integrate the networks of BRTS, metro rail, and local busway system in the city (Fig. 8.2). Ahmedabad Metro, though planned later, will be integrated with BRTS, and both BRTS and metro form an integral part of the Ahmedabad Urban Development Plan of 2021. The corridors are planned to ensure connectivity and coverage across the city. The routes for the local bus systems are now being planned for these to complement the BRTS network. BRTS is now integrated in the local and regional development plans looking at transit corridors to ensure wider coverage (AUDA 2015). The project is among the most successful public transport projects implemented on a public-private partnership model.

Despite several successful initiatives, BRTS implementation in India has experienced challenges and replication. Key challenges include lack of ownership of the system by the planning and implementing agencies, reluctance from private vehicle owners to share road space, and inadequate clarity in implementation (Mahadevia et al. 2013a). It was observed that BRTS systems fail to benefit the urban poor due to flaws in design and fare structure. While emission reductions achieved are modest, the system has potential to bring about sizable mitigation benefits. Evaluation of BRTS projects should be done in a slightly longer time frame as impacts may not be evident during initial years. Challenges of attracting private transport users, if overcome, can make the BRTS an attractive option. This will require supporting measures including pricing policies for private motorized vehicle users

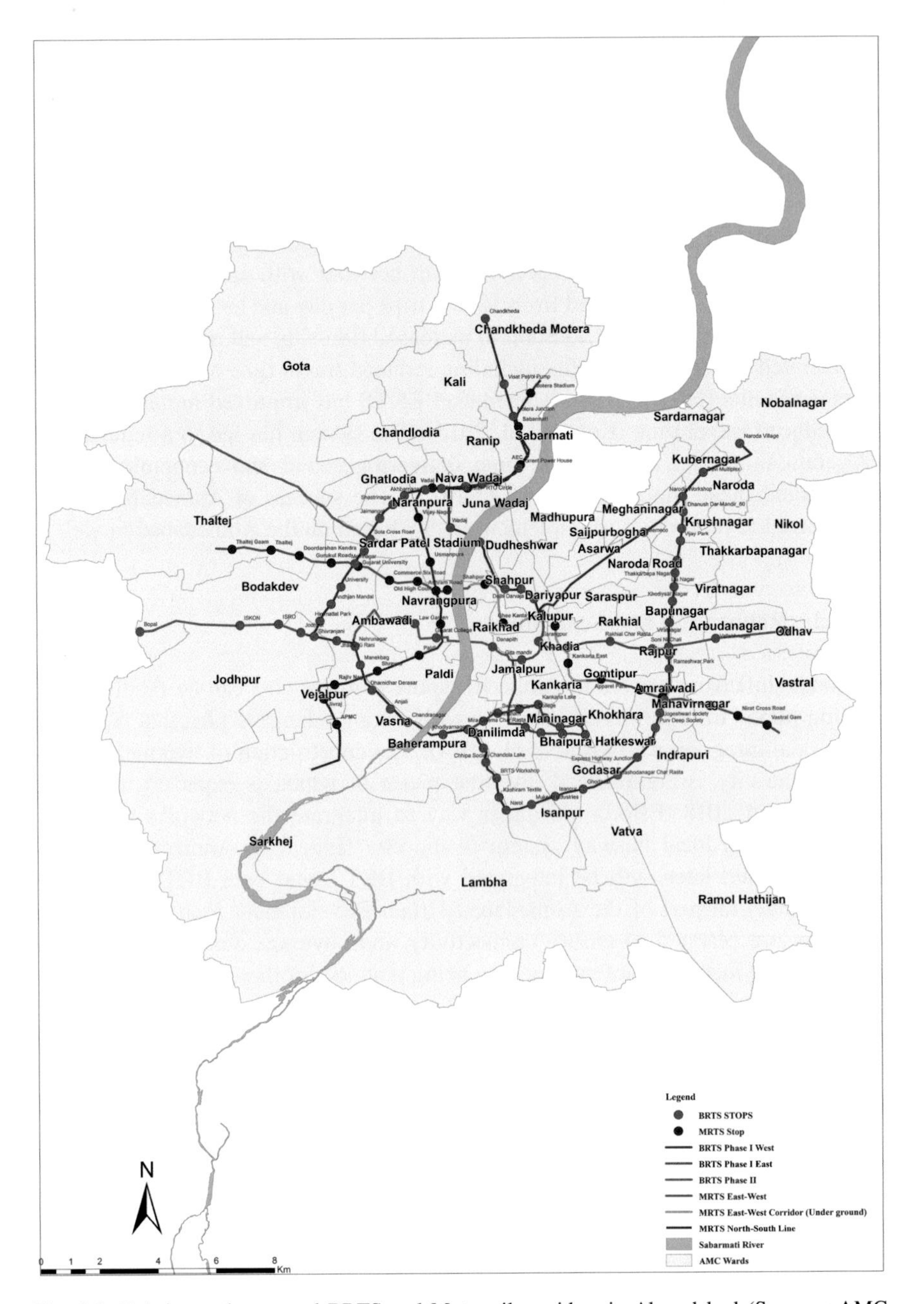

Fig. 8.2 Existing and proposed BRTS and Metrorail corridors in Ahmedabad (Sources: AMC 2012; AUDA 2015 and GoG 2014)

and supporting infrastructure including feeder systems and parking facilities at stations. For BRTS to become a best practice, efforts will have to bring in social equity by enhancing infrastructure for non-motorized transport along BRTS corridors, introduce measures to facilitate modal shift from private modes, and integrate the BRTS with other public transport systems in the city (Mahadevia et al. 2013b). Inclusiveness will depend on understanding the mobility needs and capacities of the users and integrating these into the plan. Planning of BRTS projects should be based on a comprehensive assessment to include all sections of the population, their travel patterns, and mode choices based on age, gender, and economic status.

8.3.5 Low-Carbon Comprehensive Mobility Plan Toolkit

In 2008, the Ministry of Urban Development (MoUD) released a toolkit for the preparation of a Comprehensive Mobility Plan (CMP) for cities. This was to encourage cities to prepare CMPs to fund urban transport projects under the Jawaharlal Nehru National Urban Renewal Mission (JnNURM). An assessment of submitted plans showed that these were not comprehensive in looking at social, economic, and environmental sustainability issues (GoI 2014c). Driven by concerns of climate change and sustainable development, the Ministry of Urban Development released a revised CMP toolkit in 2014. As part of a UNEP-supported project, a low-carbon comprehensive mobility plan was prepared for three cities in India—Udaipur, Rajkot, and Visakhapatnam.

The toolkit provides clear guidance on integrating land use and transport planning in order to align the transport plan with the development plans/master plans of the city. The revised toolkit incorporated learnings from implementation of earlier CMPs and from the three city case studies. The revised document was therefore more comprehensive—it incorporated elements of environmental sustainability and inclusiveness including poverty and gender issues. The revised document moves away from a deterministic approach to a more flexible scenario-based approach. This allows cities to look at the impacts of the conventional scenario where land use will follow the development plan or master plan and explore specific interventions in the alternate scenario including land use, infrastructure, public/non-motorized modes, and regulations (GoI 2014c). For instance, the Low Carbon Mobility Plan for Rajkot focuses on alternate transport scenarios based on land use, public transport and non-motorized transport, and technology-based scenarios.

8.3.5.1 Replication and Scalability

The toolkit is a comprehensive document as it integrates multiple objectives of environmental quality, inclusiveness, gender balance, and emission mitigation. This is also a unique initiative at the national level aimed to enable subnational transport plans. The process involved active stakeholder engagement by involving experts, city officials, and other stakeholders and incorporating these suggestions into the final document. The process also included a review of all existing documents and policies including government reports, reports of the expert committees and groups on urban transport of the 11th and 12th five-year plans, national missions, guidelines and codes, and global case studies on transport plans.

The methodology can highlight strategies and projects for implementation in the short term, medium term, and long term allowing cities to integrate these within the overall plans (UNEP 2015). This will generate a common and comparable database of cities on transport indicators that will further facilitate adoption by a number of cities especially smaller cities. In addition, a robust methodology will enable cities to develop proposals for funding including international climate finance.

8.3.6 *Dedicated Freight Corridor (DFC)*

The Dedicated Freight Corridor Project is initiated by the Government of India to develop transport corridors dedicated for freight transport (DFCCIL 2014). This key objective is to facilitate faster freight transport and meet market needs more effectively. In addition, the creation of the extensive infrastructure will facilitate the growth of industrial corridors and logistic parks leading to regional and national economic benefits. The primary reasons for undertaking the project of this scale were the rapidly rising demand for freight transport and the inadequacy of the existing rail infrastructure. Presently, there are plans for developing two corridors—the western DFC and eastern DFC.

The western corridor covers 1483 km between Delhi and Mumbai. The introduction of the corridor is expected to result in a major shift from road- to rail-based transport. In terms of energy implications, this will increase efficiency and reduce the demand for oil while increasing the share of electricity. By 2046, the western DFC project is expected to bring down a substantial 81 % reduction in annual CO_2 emissions compared to the "no-project scenario." With increasing decarbonization of electricity in the future, this will generate significant low-carbon benefits. This will result in a cumulative emission reduction by nearly 170 million tons of CO_2 over 30 years (Pangotra and Shukla 2012). The corridor will enhance regional connectivity, a critical input to deliver regional economic benefits.

8.3.6.1 Highlights

Historically, Indian railways had dominated the inland movement of goods. Over time, economic growth led to a significant demand for freight transport; however, rail transport infrastructure did not meet the growing demand resulting in a growing share of road transport in overall freight transport (RITES 2009). A common corridor for passenger and freight resulted in high transaction time and costs due from inefficient operations. The modal shift from rail to road is not favorable given the efficiency of rail in terms of energy and CO_2 emissions. The dedicated corridor for freight transport will deliver emission reductions from modal shift and additionally from increased efficiency of movement. In addition, India will be able to leverage global economic opportunities through better internal connectivity between centers and ports. This will facilitate industrial development along the corridor generating significant jobs in small towns and villages along the route. The case study highlights that large transport infrastructure projects have major impact on CO_2 emissions. A strong case for replication of freight corridors is the additional dimension of sustainability from simultaneous environmental and development benefits for the country.

8.3.7 National Electric Mobility Mission Plan

Recently, the Prime Minister of India launched the National Electric Mobility Mission Plan (NEMMP 2020) (GoI 2012) with a view to enhance national energy security, mitigate adverse environmental impacts of vehicle, and develop domestic manufacturing capabilities. The Plan envisions the sale of around seven million electric vehicles resulting in fuel savings of nearly 2.5 million tonnes. The NEMMP focuses on demand creation, manufacturing, R and D, and development of charging infrastructure (GoI 2012). Within these, the plan proposes phase-wise targets and strategies for implementation.

EVs could have vital implications for energy security, local air quality, GHG mitigation, and increasing renewable share in the electricity sector. It is obvious that electric vehicles will play a significant role in India's sustainable transport transition. Around the year 2000, only a couple of electric two-wheelers were available in the Indian market. However, the market has expanded, and over two dozen different two-wheelers are available in the market at present. Efforts are under way by electric vehicle manufacturers to provide options that can reduce charging time and increase awareness among consumers regarding lower fuel and maintenance costs of E4Ws compared to conventional cars.

8.3.7.1 Highlights

The NEMMP is a good starting point to give an impetus to the country's manufacturing sector, enhance research in electric vehicles, and upgrade infrastructure, all of which will be instrumental in the penetration of electric vehicles in the country. The policy sets the direction and signals to manufacturers including private players.

The NEMMP is a comprehensive policy that will facilitate green growth by enabling environmental innovation and facilitating the development of a competitive domestic market for electric vehicles, green jobs, and local air quality benefits. By laying down actions in a phase-wise manner, it sets down initial direction and sets long-term targets for scaling up.

EVs are at a relatively initial stage in India. Scaling up EV penetration in India and making these competitive vis-à-vis conventional vehicles will require financial incentives for electric vehicles, improved infrastructures for charging and other local incentives (Shukla et al. 2014). Supportive and enabling policies have the potential to increase the share of electric two-wheelers from 40 to 100 % and electric cars to 40 % and reduce oil demand by 39 Mtoe. EVs will require upfront investments; however, savings from the reduced oil demand as a result of shift to electric mobility will far exceed the support provided, thereby making this economically viable.

8.4 Conclusion

Transportation has multifarious interfaces with economic development and environment. Transport networks create access to markets and render economic efficiency. In an emerging nation like India, the demand for transport will grow through this century driven by urbanization, industrialization, and rising income. The experience of developed countries shows that the business-as-usual transport policies lead to energy-intensive and oil-dependent transport leading to high GHG emissions.

India is a geographically diverse and vast country. National transport policies are crafted keeping in view the diversity of transport demand, appropriate mix of modes, technologies, fuels, and corresponding infrastructures. The transport system architecture varies at national and subnational levels and so do policy interventions. Transport decisions interface with numerous other development policy domains, e.g., land use, energy, environment, technologies, and finance. The transport decisions have inherent long-term lock-ins lasting several decades. The transport policy making needs long-term perspective and concurrent attention to interface with multifarious development goals. Climate change is now an added interface to which transport policy makers have to pay their attention. The assessment of development policies and plans of several countries in Asia shows that their development policies were not aligned with climate change goals, though their focus on other development and environment objectives like energy security and local air pollution has led to reduced GHG emissions (ADB 2012). For India, the studies have shown opportunities to align policies to simultaneously ensue multiple development and climate objectives (Menon-Choudhury et al. 2007). This chapter presents selected best practices that have shown the promise of gaining multiple co-benefits which are scalable and replicable (Table 8.3). The case studies also show that governance system, including monitoring, reporting, evaluation, and correction, is vital for ensuring replication and scalability.

The case studies represent best practices related to policies and projects. Evidently, the challenge is to replicate and scale up these practices to gain sizable CO_2 mitigation together with co-benefits vis-à-vis various national sustainable development goals. The global mitigation agreements now provide the opportunity to

Table 8.3 Summary assessment of best practice case studies

Sector/ scale	Measures for direct CO_2 benefits	Sustainability Co-benefits	Elements of replicability and scalability to gain sustainability and low-carbon benefits
Policy-related case studies			
CMP Toolkit *Focus*: urban mobility *Scale*: national, city	Integrates transport modes with land use Fuel/technology mandate Measures to reduce transport demand	Lower cost to consumer Lower congestion Improved air quality Safety and security Inclusive and affordable transport	Proactive and continuous stakeholder engagement Integration of climate, inclusiveness, environment, and quality of life Methodology for comparable and consistent databases across cities Cities have flexibility to tailor context-specific interventions Enables capacity building and early interventions in smaller cities
NEMM Plan *Focus*: EVs *Scale*: national, city	Electric vehicles using low-carbon electricity can reduce CO_2 sizably	Local air quality benefits Energy security co-benefits Reduced noise Batteries help electric load management and backup power	Charging infrastructure is easy to replicate and upscale Two-wheeler manufacturing is easy and is readily replicated Subnational policy makers have incentive to attract EV manufacturers, and this can help replication and scalability
Auto Fuel Policy *Focus*: clean air/fuel *Scale*: national, city	Direct GHG benefits can accrue from fuel economy standards GHG co-benefits from local emissions targets	Local air quality co-benefits National energy security co-benefits	AFP 2003 helped replicate and upscale Auto Fuel Policy Vision 2025 AFP process was stakeholder intensive which helped in acceptance and replication AFP process is amenable for conjoint mitigation of local pollutants and GHG emissions
Project-related case studies			
Delhi Metro *Focus*: mass transit in cities *Scale*: city	CO_2 emission reduction due to modal shift from motorized transport Low-carbon electricity in the future can deliver sizable CO_2 mitigation	Air quality co-benefits Reduced congestion Improved connectivity Efficient technology enhances energy security	Scalability in Delhi and replication in a number of cities has happened Incremental carbon finance is expected to help replication under NAMAs framework Delhi Metro co-benefits

(continued)

Table 8.3 (continued)

Sector/ scale	Measures for direct CO_2 benefits	Sustainability Co-benefits	Elements of replicability and scalability to gain sustainability and low-carbon benefits
			assessment provides methodology and benchmarks for replication and scalability elsewhere
BRTS *Focus*: mass transit *Scale*: city	Modal shift delivers CO_2 mitigation System efficiency delivers CO_2 mitigation	Air quality co-benefits Saves time due to rapid mobility Higher safety from dedicated lanes	Replicated in several other cities in India Better integration with other public transport systems, into the city development plan, and within the larger regional context PPP model is suitable for BRTS and it is replicable an scalable
Dedicated freight corridor *Sector*: freight *Scale*: regional	CO_2 savings due to shift to efficient rail mode away from road transport Low-carbon electric traction will mitigate CO_2 in low carbon future	Facilitates development of industrial parks along the corridor leading to regional and national economic benefits Promotes industrialization along the corridor	Comprehensive view of economic and environment benefits Improve efficiency of freight movement Scaling up to facilitate national economic development and deliver sizable mitigation benefits Integration of coastal and landlocked areas which is very essential for balanced national development and global integration

leverage additional funding from climate finance instruments like Green Climate Fund as well as Nationally Appropriate Mitigation Actions (NAMAs). The additional funds can be the lever for fast-track replication and upscaling of current best practices. Globally, cities have proposed projects under NAMAs that include implementation of low-carbon mobility plans and demand management including road pricing, parking policies, investing in mass transit, and increasing the share of non-motorized and public transport.

The overarching vision of the case studies in the chapter is largely focused on aligning national transport policies in line with the global target of 2 °C temperature stabilization by the end of the century. However, given the climate risks to infrastructure projects, the protection of transport assets from the future climate change is one of the areas where more attention is needed. While climate risks are not formally factored into the existing transport policies and projects, the methodologies to identify and mitigate major climate risks to the transport projects by improved design and construction methods is gaining attention.

The case studies presented in the paper represent just a few of the promising interventions. There are equally promising initiatives such as investing in non-motorized transport. Recently, India's Ministry of Urban Development has released the bicycle sharing toolkit to promote non-motorized transport in cities. Several cities including Ahmedabad, Delhi, Vishakhapatnam, and Chennai have initiated construction of infrastructure for non-motorized transport and cycle-sharing schemes. This is an important focus area as it can deliver multiple gains of mobility, safety, emission reductions, and social inclusion.

The "best practice" assessment presented in this chapter shows promise of delivering multiple objectives and the possibility of replication and upscaling. The policies and projects represented in the case studies show that urbanization is the key driver of future transport system choices in India. Rational transport system therefore needs integration of inter- and intracity transport choices. The lessons from these studies provide important learnings for designing policies and projects elsewhere. The assessments of case studies show that successful implementation of national policies at the subnational level requires widely agreed goals and targets and support from the national government. The support can be in the form of capacity building, technology, or finance. Overall, the chapter argues for (1) integrating transport policies with local, national, and global objectives, (2) a comprehensive assessment of the impacts (co-benefits and risks) of policies and project from the planning to the post-implementation stage, and (3) cooperation and knowledge sharing among cities and regions facilitated by the national government for cross-learning and transfer of best practices.

References

ADB (2012) Policies and practices for low carbon green growth in Asia. Asian Development Bank Institute, Philippines

AMC (2012) Ahmedabad Municipal Corporation. www.egovamc.com. Accessed 1 Mar 2015

AUDA (2015) Ahmedabad Urban Development Authority. www.auda.org.in. Accessed 1 Mar 2015

BEE (2014) Bureau of energy efficiency. http://www.beeindia.in/. Accessed 14 Feb 2015

Bubeck S, Tomaschek, Ulrich Fahl J (2014) Potential for mitigating greenhouse gases through expanding public transport services: a case study for Gauteng Province, South Africa. Transp Res D 32:57–69

CPCB (2008) Status of the Vehicular pollution Control programme in India, 2008 http://www.cpcb.nic.in/upload/NewItems/NewItem_157_VPC_REPORT.pdf. Accessed 3 Mar 2015

CPCB (2012) National ambient air quality status & trends in India-2010. Central Pollution Control Board, Ministry of Environment and Forests, New Delhi

Creutzig F, He D (2009) Climate change mitigation and co-benefits of feasible transport demand policies in Beijing. Transp Res Part D: Transp Environ 14(2):120–131. doi:10.1016/j.trd.2008.11.007

CRRI (2011) Central Road Research Institute. http://crridom.gov.in/
DFCCIL (2014) Dedicated Freight Corridor Corporation of India. http://www.dfccil.gov.in/dfccil_app/home.jsp. Accessed 10 Jan 2015
Dhar S, Shukla PR (2014) Low carbon scenarios for transport in India: co-benefits analysis. Energ Policy. doi:http://dx.doi.org/10.1016/j.enpol.2014.11.026i
Dhar S, Pathak M, Shukla PR (2013) Low carbon city: a guidebook for city planners and practitioners. UNEPRISO Centre on Energy, Climate and Sustainable Development, Denmark Technical University, ISBN: 978-87-92706-27-0
DMRC (2014a) Delhi Metro Rail Corporation. http://www.delhimetrorail.com/press_reldetails.aspx?id=746xECETA6Qlld
DMRC (2014b) Annual report 2013. Delhi Metro Rail Corporation. Delhi Metro Rail Corporation Limited. New Delhi
DMRC (2015) Delhi Metro Rail Corporation Ltd. http://www.delhimetrorail.com/funding.aspx. Accessed 15 Feb 2015
Garg A, Naswa P, Shukla PR (2013) Impact assessment and management: framework for infrastructure assets: a case of Konkan railways. UNEP Risø Centre, Roskilde
GGBP (2014) Green growth in practice. Global Green Growth Institute. www.ggbp.org
Goel D, Gupta S (2014) The effect of metro rail on air pollution in Delhi, Working paper No. 229. Center for Development Economics
GoG (2014). Detailed project report for Ahmedabad Metro Rail Project (Phase – I). Government of Gujarat. http://www.gujaratmetrorail.com/mega.html. Accessed 6 Dec 2014
GoI (2003) Auto fuel vision and policy 2003. Government of India, New Delhi
GoI (2008) Prime Minister's council on climate change: national action plan on climate change. Government of India, New Delhi
GoI (2012) National electric mobility mission plan 2020. Department of Heavy Industries, Ministry of Heavy Industry and Public Enterprises, Government of India, New Delhi
GoI (2014a) High speed rail corporation of India Limited. http://hsrc.in/ Accessed 31 Oct 2014
GoI (2014b) Auto fuel vision and policy 2025. Government of India, New Delhi
GoI (2014c) Preparing a comprehensive mobility plan (CMP)- a toolkit ministry of urban development. Government of India, New Delhi, September
GoI (2015) Indian railways – a white paper. http://www.indianrailways.gov.in/railwayboard/uploads/directorate/finance_budget/Budget_2015-16/White_Paper-_English.pdf. Accessed 15 May 2015
ICCT (2013a) Overview of India's vehicle emissions control program past successes and future prospects. International Council on Clean Transportation, Washington, DC
ICCT (2013b) The case for early implementation of stricter fuel quality and vehicle emission standards in India. International Council on Clean Transportation, Washington, DC
IEA (2012) World energy outlook. International Energy Agency, Paris
IGES (2011) Transport co-benefits approach a guide to evaluating transport projects. Institute for Global Environmental Strategies, Japan
Khanna P, Jain S, Sharma P, Mishra S (2011) Impact of increasing mass transit share on energy use and emissions from transport sector for National Capital Territory of Delhi. Transp Res Part D: Transp Environ 16(1):65–72
Mahadevia D, Joshi R, Datey (2013a) A low-carbon mobility in India and the challenges of social inclusion Bus Rapid Transit (BRT) Case Studies in India. UNEP Risø Centre on Energy, Climate and Sustainable Development. Technical University of Denmark, ISBN: 978-87-92706-77-5
Mahadevia D, Joshi R, Datey (2013b) Ahmedabad's BRT system: a sustainable urban transport panacea? Econ Polit Weekly 48:56–64
Matsumoto T, Nuttal C, Bathan G, Gouldson A, Pathak M, Robert A, Welch D (2014) National-subnational integration for green growth. Chapter 8. In: Green growth in practice: lessons from country experiences. Green Growth Best Practice. http://gggi.org/wp-content/uploads/2014/03/Green-Growth-in-Practice-062014-Full.pdf. Accessed September 25, 2015

Menon-Choudhury D, Shukla PR, Hourcade JC, Mathy S (2007). Aligning development, air quality and climate change policies for multiple dividends. CSH Occasional Paper NO 21, Publication of the French Research Institutes in India, December

MoEF (2010) India: green house gas emissions 2007, Indian Network for Climate Change Assessment (INCCA), Ministry of Environment and Forests (MoEF). Government of India, New Delhi. Accessed 13 Sept 2013

MoUD (2006) National urban transport policy. Ministry of Urban Development. http://moud.gov.in/sites/upload_files/moud/files/pdf/TransportPolicy.pdf. Accessed 30 Jan 2014

Murty MN, Dhavala K, Ghosh M, Singh R (2006) Social cost-benefit analysis of Delhi Metro. Institute of Economic Growth, Delhi, MPRA paper no. 1658 http://mpra.ub.uni-muenchen.de/1658/1/MPRA_paper_1658.pdf. Accessed 15 Jan 2014

NIUA (2014) Urban transport initiatives in India: best practices in PPP. National Institute of Urban Affairs. http://carbonn.org/uploads/tx_carbonndata/File1_AHMEDABAD%20BRTS.pdf. Accessed 31 Jan 2015

Olsen KH (2013) Sustainable development impacts of NAMAs: an integrated approach to assessment of co-benefits based on experience with the CDM Low Carbon Development. Working paper no. 11, UNEP Riso Centre, Denmark Technical University, Denmark

Pangotra P, Shukla PR (2012) Infrastructure for low-carbon transport in India: a case study of the Delhi-Mumbai Dedicated Freight Corridor. ISBN: 978-87-92706-69-0

Pathak, M., Shukla P.R., Garg A. and Dholakia H. (2015). Integrating Climate Change in City Planning: Framework and Case Studies Ch. 8 In: Cities and Sustainability: Issues and Strategic Pathways. Mahendra Dev S.,Sudhakar Yedla S. (Eds). Springer Proceedings in Business and Economics ISBN:978-81-322-2310-8

RITES (2009) Total transport system study on traffic flows and modal costs. Planning Commission. Government of India

Rogat J, Dhar S, Joshi R, Mahadevia D, Mendoza JC (2015) Sustainable transport: BRT experiences from Mexico and India. WIREs. Energy Environ. doi:10.1002/wene.162

Sharma N, Singh A, Dhyani R, Gaur S (2014) Emission reduction from MRTS projects – a case study of Delhi metro. Atmos Pollut Res 5:721–728

Shukla PR, Dhar S, Pathak M, Bhaskar K (2014) Electric vehicle scenarios for India. UNEP DTU Partnership, Centre on Energy, Climate and Sustainable Development Technical University of Denmark. ISBN: 978-87-93130-22-7

Tiwari G (2011) Metro systems in India: case study DMRC. Promoting Low Carbon Transport in India. National workshop 19-20 October, 2011, Delhi

UN (2014) World urbanization prospects: the 2014 revision. Department of Economic and Social Affairs. Population Division. United Nations, New York, 2014

UNEP (2015) Low-carbon comprehensive mobility plan for Rajkot. www.unep.org/Transport/lowcarbon/Pdf's/Rajkot_lct_mobility.pdf. Accessed 15 Feb 2015

UNFCCC (2014) Metro Delhi, India. United Nations Framework Convention on Climate Change. https://cdm.unfccc.int/Projects/DB/SQS1297089762.41/view. Accessed 14 Feb 2015

UNFCCC (2015) Ahmedabad bus rapid transport system. http://unfccc.int/secretariat/momentum_for_change/items/7098.php. Accessed 31 Jan 2015

Vesely A (2011) Theory and methodology of best practice research: a critical review of the current state. Cent Eur J Public Policy 5(2):99–117

Chapter 9
Potential of Reducing GHG Emission from REDD+ Activities in Indonesia

Rizaldi Boer

Abstract Loss of forest cover in large scale in tropical region will have impact on climate significantly. This will change air pressure distribution and shift the typical global circulation patterns and change rainfall distribution. Its contribution to the increase of greenhouse gas emission will also enhance global warming and may increase the frequency and intensity of extreme climate events. Deforestation in the three tropical regions, Amazon, Central Africa, and Southeast Asia, still continues. Without significant change in forest protection efforts, the loss of forests in these three regions by 2050 will reach about 29, 98, and 44 %, respectively.

Indonesia has the largest tropical forest in SEA; the contribution of emission from land use change and forest (LUCF) reached 60 % of the total national emission, much higher than energy sector. During the period 1990–2013, the total loss of natural forest reached about 19.7 million hectares or about 0.822 million ha per year. Without significant change in forest protection program, within the period 2010–2050, Indonesia may lose 43.4 million ha of forest or equivalent to deforestation rate of 1.08 million ha per year. Potential of reducing emission from REDD+ activities is quite big. By increasing expenses of the government by 1 % annually on top of the external investment for technology change, without necessity of direct forest protection (e.g., increasing agriculture productivity reduces pressure on forests), the deforestation rate could reduce to about 0.337 million ha per year.

The issuance of innovative financing and incentive policies for improving land and forest management may further increase the potential of reducing emission from REDD+ activities. Some of the policies include the use of debt-for-nature swap (DNS) scheme for accelerating the development of forest management units in open access area, incentive for permit holders for accelerating the development of timber plantation on degraded land, and increase community access to fund for green investment. The incentive system for the permit holders is for handling land tenurial issues or conflicts. The incentive could be in the form of reducing or exemption of administration/retribution fees for certain period of time depending on the level of conflicts. Policy allowing for transferring the funds to a financing system is relatively easy to be accessed by the community such as blending

R. Boer (✉)
Centre for Climate Risk and Opportunity Management in Southeast Asia and Pacific, Bogor Agricultural University; CCROM SEAP-IPB, Bogor, Indonesia
e-mail: rizaldiboer@gmail.com

S. Nishioka (ed.), *Enabling Asia to Stabilise the Climate*,
DOI 10.1007/978-981-287-826-7_9

financing, a financing system that synergizes all financial sources such as CSR funding; government funding such as state budget (APBN); and local government budget (APBD) funds, banking, and international funding. This system can help leverage private funding and supports regional development by supporting community activities in urban agriculture and agroforestry including building human resource capacity through assistance and training activities.

Keywords Extreme weather/climate events • Tropical forest • Greenhouse gas emissions • REDD+ activities • Financing policies • Incentive and disincentive policies

Key Message to Policy Makers

- Loss of large scale of tropical forest will bring more and intensify extreme weather/climate events.
- Loss of forest in Indonesia 1990–2013 accounts for most of deforestation in Southeast Asia with average loss of about 0.822 Mha per year.
- From the period 2010–2050, Indonesia potentially can reduce its deforestation rate more than half of the current rate to 0.337 Mha per year.
- The potential reduction of the deforestation may be achieved by facilitating changes in technologies without necessity of direct forest protection.
- Implementation of innovative financing policies and incentive/disincentive system may further reduce emission from REDD+ activities.
- The payment from REDD+ activities might offset the government additional expenses incurred in facilitating the changes.

9.1 Introduction

Forest plays a significant role in regulating our climate. Regional climates were sensitive to change of types and density of vegetation (Dickinson and Henderson-Sellers 1988; Shukla et al. 1990; Dale 1997; Avisar and Werth 2005). Loss of forest cover in large scale directly alters the reflectance of the earth's surface, induces local warming or cooling, and finally changes air pressure distribution. The changes in air pressure distribution shift the typical global circulation patterns and change rainfall distribution. At present, deforestation of tropical regions continues at high rate (Houghton et al. 2012). The major impact of tropical deforestation on precipitation may occur in and near the deforested regions themselves. However, a strong impact will be propagated by teleconnections along the equatorial regions and to mid-latitudes and even high latitudes even though not as strong as in the low latitude. Based on climate modeling analysis, deforestation of tropical regions (Amazon, Central Africa, and Southeast Asia) significantly affects precipitation at mid- and high latitudes through hydrometeorological teleconnections (Avisar and Werth 2005). Without significant change in forest protection efforts, the loss of forests in these three regions by 2050 will reach about 29, 98, and 44 %, respectively (Schmitz et al. 2014).

Deforestation will also contribute to the increase of GHG emission to the atmosphere. In the long term, the increasing GHG concentration in the atmosphere will cause an increase in global temperature and global climate. New finding from the 5th AR of IPCC indicated that agriculture, forest, and other land uses represent 20–24 % of global emission. Without mitigation efforts, the contribution of this sector may increase to 30 % by 2030. The three tropical regions, South America (TSAm), Southeast Asia (SEA), and tropical Africa (Af), are the main contributors to the global emission from land use change and forestry (Fig. 9.1). In the last 50 years, the rate of the emission from this sector tended to increase, except in South America (Houghton et al. 2012), and it is the largest and most variable single contributor to the emission from land use change (Le Quere et al. 2013). It is clear that deforestation in the short term will affect the regional climate and in the long term enhances global warming causing the increase in frequency and intensity of extreme weather and climate events.

Among Southeast Asian countries, Indonesia has the largest forest area. Rate of deforestation fluctuates from year to year; however, in general it tended to increase. GHG emission from land use change and forest (LUCF) has been found to be the major contributor to the total national emission. It accounted for about 60 % of the total national emission, much higher than energy sector (MoE 2010). Efforts for reducing national emission have been prioritized on this sector (Bappenas 2010). Potential of reducing emission from REDD+ activities, i.e., reducing deforestation and forest degradation, maintaining role of forest conservation, implementing

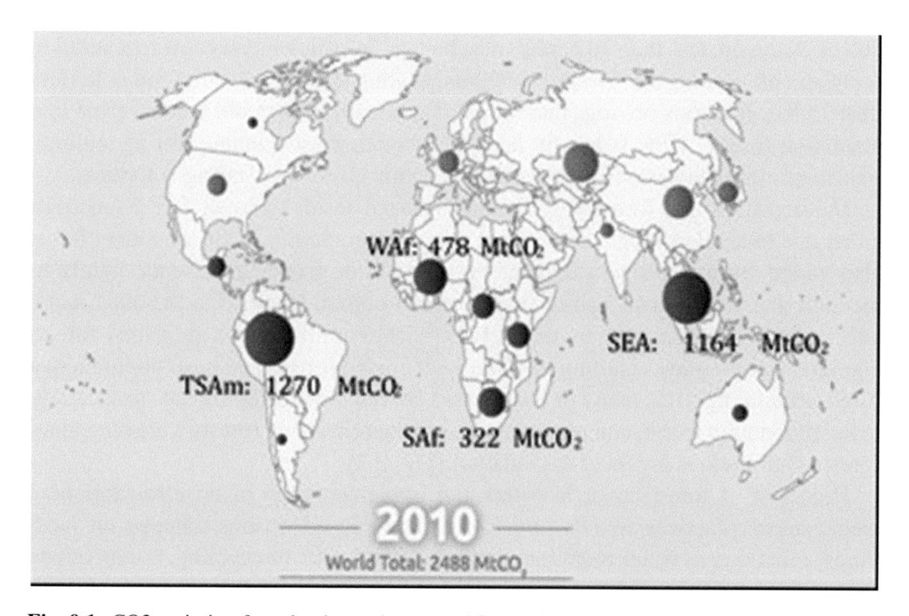

Fig. 9.1 CO2 emission from land use change and forest (http://www.globalcarbonatlas.org)

sustainable forest management, and enhancing forest carbon sequestration, is quite big. Potentially the LUCF sector can become net sink by 2030 (MoE 2010). However, a number of innovative policies are required to realize this.

9.2 Indonesian Forest

In 2013, Indonesia has forest area of about 128.4 million ha and 59.4 million ha of non-forest area (APL; MoFor 2014). By function, the forest area is classified into 5 categories, namely, conservation forest (HK), protection forest (HL), limited production forest (HPT), production forest (HP), and convertible production forest (HPK). Conservation forest is designated for conservation purposes (Act No. 5/1990, Sanctuary Reserve area, Nature conservation, and Game Hunting Park), while protection forest to serve life support system, maintain hydrological system, prevent of flood, erosion control, and seawater intrusion, and maintain soil fertility. Production forest is aimed for timber and non-timber production, while convertible production forest (HPK) is for non-forest-based activities such as agriculture, settlement, etc. Thus, this forest can be released to become a non-forest area (APL).

Referring to its function, forest clearing and conversion of forest land in HK and HL to other land uses are not allowed. Deforestations occurred in these forests mostly from illegal activities such as logging, forest encroachments, and forest fires. On the contrary, forest clearings are permitted within HP and HPT, especially over unproductive forested areas for the purpose of establishing timber plantation. Unproductive forests comprised of forest areas with less than 25 core tress/ha with dbh of 20 cm up, less than 10 parent trees/ha, and insufficient/very few regeneration (numbers of seedling are less than 1000/ha, sapling less than 240/ha, and poles less than 75/ha). It is thus obvious that not all degraded forests could be converted into plantation forests. HPK is legally designed for other uses, mainly for agriculture, transmigration, plantations, and settlements, thus all forest clearing activities.

Deforestation and forest degradation occurred in all types of forest functions either due to legal or illegal activities. Level of degradation of the secondary forest also varied from heavily to lightly degraded. With proper treatments, lightly to medium degraded forests can recover to reach climax forests. On the other hand, due to improper management and less strict law enforcement, degraded forests continue deteriorating resulting in severely degraded forests and meet unproductive forest criteria. In 2012, many of forest areas are not covered by forests, particularly in the production forest, and more than half of the remaining forests were secondary forests with various levels of degradation (Fig. 9.2).

High lost of forest cover in forest and implementation of unsustainable land management practices in non-forest area also caused serious damage on land. Forest functions as water retention, erosion control, nutrient cycling, microclimate regulator, and carbon retention were completely depleted. Many of the lands in both forest and non-forest areas are critical. Based on the level of damage, the critical

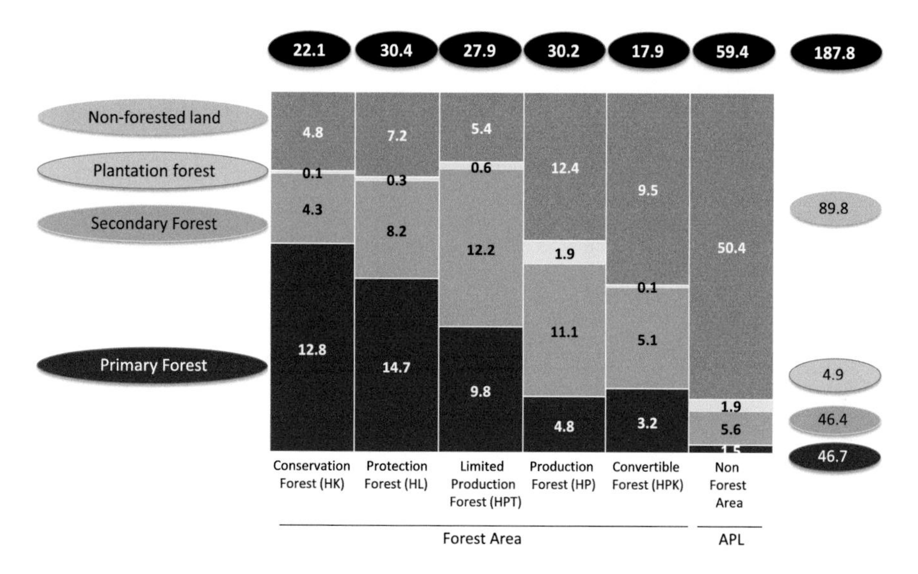

Fig. 9.2 Area and forest condition in 2013 (MoFor 2014)

lands are classified into five categories, i.e., very critical, critical, rather critical, potentially critical, and not critical. In 2011, the total area of critical lands had reached 27.3 million ha, comprised of 22.0 million ha critical and 5.3 million very critical (MoFor 2014). The critical and very critical lands have been prioritized for the implementation of land rehabilitation program.

9.3 Deforestation and Trend of CO2 Emission

Factors causing deforestation and degradation varied among islands. In Sumatra in the early 1980s, the main driver of deforestation was the establishment of settlement through transmigration program, while in Kalimantan, it was mainly due to excessive timber harvesting (MoE 2003). Logging is not responsible for the deforestation of Indonesian forests. However, road network systems that have been developed during timber harvesting have opened the access of community to the forest area. Attractiveness of timber products, high agriculture income, and open access market have increased the insecurity of the forest. Combination of high logging extraction coupled with community encroachment has caused high rates of forest degradation and deforestation.

Based on recent data published by the Ministry of Environment and Forestry (MEF) during the period 1990–2013, the total loss of natural forest reached about 19.7 million hectares or about 0.822 million ha per year. The rate of the deforestation quite varied between periods (Fig. 9.3). The highest rate occurred during the period of 1996–2000 and the lowest in the period 2009–2012. The highest period

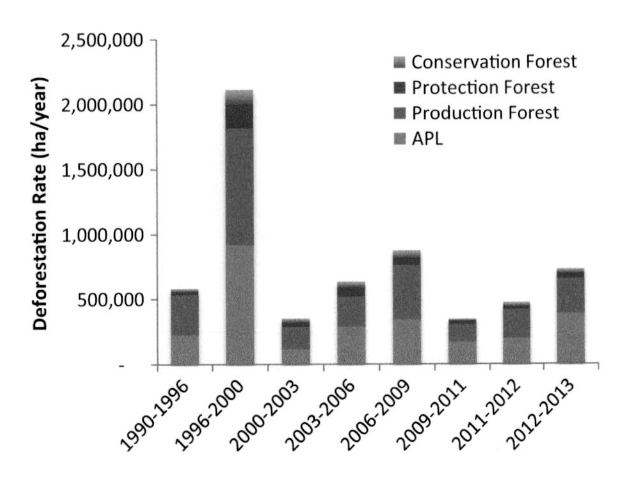

Fig. 9.3 Rate of deforestation in Indonesia between 1990 and 2013 (Directorate of Forest Resource Inventory and Monitoring 2015)

occurred during government transition period between new order ("*orde baru*") to reform ("*reformasi*") government.

Further analysis to land cover data of 1990–2013 showed that cropland conversion was found to be one of the key drivers causing deforestation (both commercial and subsistence agriculture). The loss of forest for the establishment of forest plantation, including expansion of settlement and other lands, is also quite significant even not as large as cropland (Fig. 9.4). Nevertheless, the area of grassland (including shrubs) also increased quite significantly during the period. This indicated that conversion of forest was not always used for meeting the land demand for development (for productive uses), but some were left as nonproductive lands. The data suggest that about half of the conversion of forest to non-forest lands ends up to grassland (including shrubs). In addition, the conversion of forest in the peatland for other uses tended to increase recently, particularly for the expansion of cropland, other lands, and establishment of timber plantation (Fig. 9.4). The rate of forest loss in the peatland is relatively higher than that in the mineral soils (Fig. 9.5).

The removal of biomass at the time of deforestation and forest degradation during the period of 1990–2013 was responsible for emission of about 0.693 Gt CO_2 per year.[1] Emission from peat decomposition of the forest lands deforested and degraded since 1990 reached about 0.115 Gt CO_2 per year (Fig. 9.6). Thus, in total the average emission due to deforestation and forest degradation occurred from 1990 to 2013 was 0.807 Gt CO_2/year. Busch et al. (2015) estimated that the average CO_2 emission from deforestation and peat decomposition in the period of between 2000 and 2010 was about 0.859 Gt CO_2 per year. Compared to this analysis, the rate of emission from deforestation and forest degradation during this period was about 0.884 Gt CO_2 per year.

[1] Stock carbons of primary and secondary forest were about 156 and 126 tC/ha respectively. The assumption was that all the removed biomass are emitted at the time of deforestation which is called as potential emission.

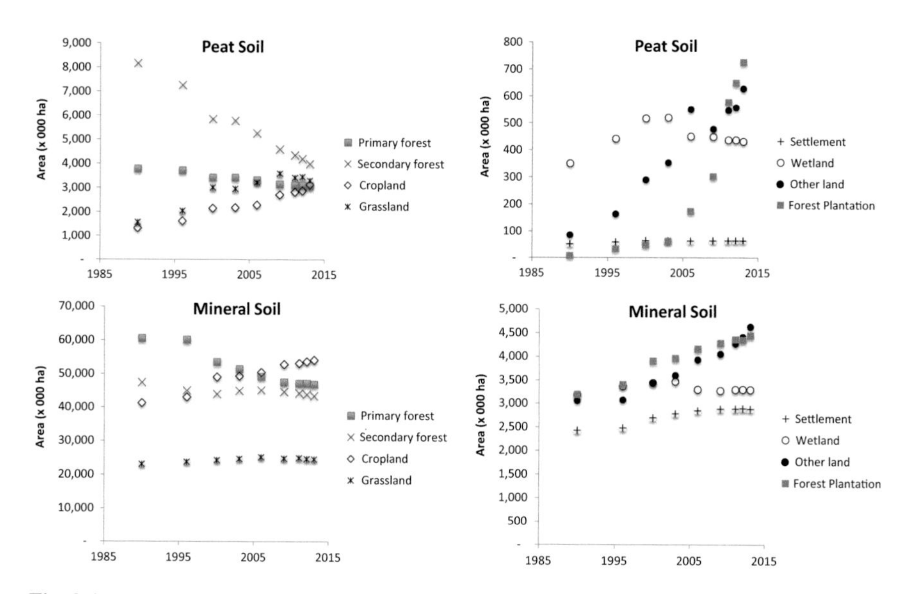

Fig. 9.4 Changes of forest land and non-forest lands in peat and mineral soils from 1990 to 2013 (Based on data from Directorate of Forest Resource Inventory and Monitoring 2015)

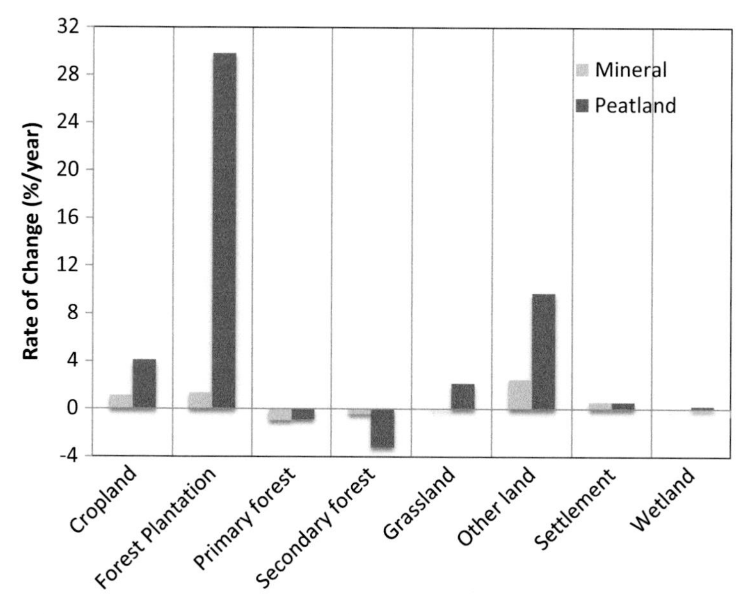

Fig. 9.5 Annual rate of change of forest lands and non-forest land areas in the period of 1990 to 2013 (Based on data from Directorate of Forest Resource Inventory and Monitoring 2015)

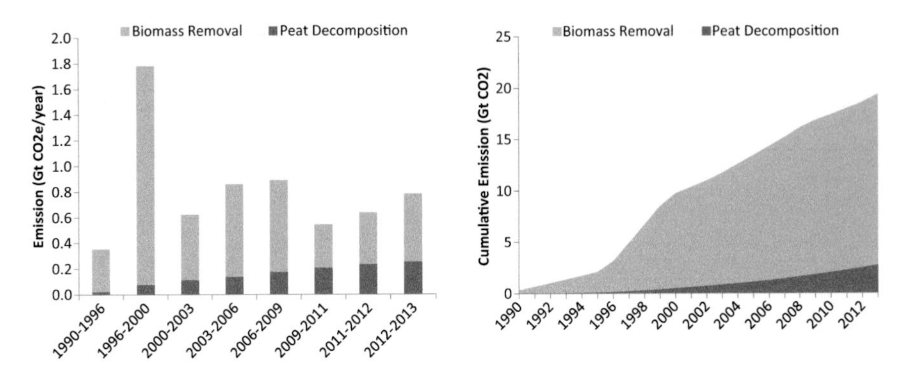

Fig. 9.6 Estimated gross CO_2 emission from deforestation and forest degradation from 1990–2013 (Emission was from biomass removal at the time of deforestation and forest degradation (biomass loss from change of forest state from primary to secondary forest) and from peat decomposition (occurred from secondary forest, timber plantation, cropland and grassland and emission factor was taken from Hergoualc'h and Verchot, (2014). The carbon sequestration after the deforestation was not taken into account.)

9.4 Low-Carbon Policies on Forest and Land Use Sector

The Government of Indonesia has issued national policies and action plan for reducing emission from land use change and forestry defined in the Presidential Regulation Number 61/2011. In general there are four main policies and actions toward low carbon (Boer 2012). *First* is accelerating establishment of forest management unit (FMU) to improve the management of land and forest resources in all forest areas. *Second* is pushing adoption of sustainable management practices in production forests by implementing mandatory forest certification systems. *Third* is reducing dependency on natural forests in meeting wood demands through acceleration of establishment of timber plantation on community lands and state lands and enhancement of sink through restoration of production forests ecosystem and land rehabilitation. *Fourth* is reducing pressure on natural forest through optimization of the use of land and improvement of land productivity. To support the implementation of these policies and actions, it is crucial to develop financing/incentive policies and development of financing system that can support their adoption and implementation by related stakeholders.

9.4.1 Forest Management Units (FMUs)

Key factors driving deforestation in Indonesia might originate from forestry sector and also from outside the forestry sector. These factors intermingle in complex processes, which are difficult to separate, which includes long drought period and characteristics of land that are rich in mineral resources but susceptible to fire interlink with management practices as well as political decision and economical considerations in the allocation of land uses, its utilizations and enforcement of rules. They both intend to pursue the goal of national development in forms of economical growth, political stability, as well as social equity and ecological sustainability. It is difficult to identify which key driver comes first and further stimulates the emergences of others. Some key drivers observed from current practices and have consequences on land use and land cover changes are forest fire, logging, timber plantation, agriculture expansion, mining, and political administration expansion.

Establishment of forest management unit (FMU) at site level has been considered as a prioritized program for improving management of forest resources and controlling deforestation and forest degradation. Urgency of FMU development especially outside Java[2] is driven by the fact that (Nugroho et al. 2011):

1. Intensive management of forest resource at site level is required as mandated by Act No. 41 Year 1999 on Forestry which states that "All forests within the territory of the Republic of Indonesia, including natural resources contained therein is controlled by the State for the greatest prosperity of the people".
2. Management of forest resources given to the private sector through the licensing mechanism for forest (IUPHH) has limited time, and when it is over, the forest area becomes unmanaged. In addition, nature of the transfer of rights to holders of the license also required close monitoring from government over the behavior of the license holders.
3. Many of investments for land and forest rehabilitation implemented in forest area (GERHAN) often fail as due to the absence of manager in the site who will manage the maintenance of the planted trees.
4. Programs for giving access to public in playing active role in managing forest resources such as community-based plantation forest (HTR), village forest, and community forest (HKm) are slowly realized, due to the absence of companion at the implementation level.

Duties and functions of the FMU (*PP. 6/2007 jo PP. 3/2008*) include (1) implementing management of forest resources which includes forest arrangement and management plan, utilization of forest area and resources, rehabilitation and reclamation of forest area, and protection and conservation of forest area; (2) translating national, provincial, and district/city forest policy to be implemented at site level; (3) implementing forest management activities in the region starting from planning, organizing, implementing, and monitoring and control; and (4) implementing the monitoring and the assessment of implementation forest management activities in its territory and opening investment opportunities to support the achievement of forest management objectives.

[2] FMU had already existed long before in all forest area in Java under the management of State Forest Company Perum Perhutani and called KPH (*Kesatuan Pemangku Hutan*).

FMU is targeted to be developed 600 units throughout forest area, and by February 2014, only 120 units were established. However, operationalization of these first 120 units remains problematic (Nugroho et al. 2014). Some of the problems include:

1. Scope and authorities of FMU in managing forest area. FMU authority is actually very powerful, but this is supported by a number of different regulations, not summarized in one single regulation. So KPH management team is not functioning optimally. As an example, annual working plan of concession holder (RKT) should be approved via KPH once a respective area has established its KPH. Articles 71–78 of government law No. 6/2007 regulate this issue. However, none of RKT now is submitted to KPH. Its function on monitoring and evaluation of concessions does not work. Therefore, it is necessary to compile a list authority of KPH mandated by regulations and laws and issue a strategic regulation on this.
2. Capacity of stakeholders and supports from local government (Province/district) are still diverse. Dynamics of local politics also very much influence their commitments in running FMU.
3. Sectoral ego does exist. There is a doubt that some functions of forestry-related agencies will be taken over by FMU.
4. Regulation No. 23/2014 about local government authority on forestry issue (incl. KPHP and FMU for protection forest (KPHL)) results a concurrent between central government and provincial government (Article 14 (1)). The regulation also indicates less role of district government. However, sites are located within administrative authority of district government.
5. Many FMUs have been legalized by MoF decree, yet do not have any organization at site level (8 out of 120 units).
6. Barriers in regulating concessions incl. RHL and issue of coordination between FMU and concession holders.
7. Independence of FMU needs to be improved. A number of regulations such as No. 61/2007 about technical guidance of general service budget management (PPK – BLUD).
8. Lack of human resources and funding.
9. Need a synchronized policy and coordination among Echelon 1 at ministry of forestry to support operationalization of FMU.
10. Socialization of FMU development has been concentrated to forestry agency at provincial level. While communication on FMU policies by central government has not touched strategic decision making at local level.
11. Mechanism on national budget is not flexible for supporting FMU.
12. Land tenure conflicts as a consequence of non-FMU area rights. Local community often claims those areas. Ministry of forestry has very weak power on this type of areas.
13. Lack of leadership and entrepreneurship in FMU directors/heads.

As mentioned above (Problem No. 7), FMU independence is needed because often local government, i.e., majorly ask for benefits – specifically financial benefits

Table 9.1 Budget management at BLUD and non-BLUD working units

	Description	Non-BLUD	BLUD
1	Administrator	Civil servant (PNS)	PNS and professional non-PNS
2	Tariff of service	Based on fair/proper norms	Cost per service unit
3	Medium-term document	Medium-term development plan (*Rencana Pembangunan Jangka Menengah – RPJM*)	Business strategic plan (*Rencana Strategis Bisnis, RSB*)
4	Budgeting	Budget work plan (*Rencana Kerja Anggaran – RKA*)	Business budget plan (*Rencana Bisnis Anggaran – RBA*)
5	Budget allocation	After DIPA (national budget) is authorized	Independent from national budget
6	Financial activities	Petty cash and all other financial activities are conducted via PEMDA bank account	Independent and has its own bank account
7	Income	Transferred to state account	Usable for operational
8	Surplus	Transferred to state account	Usable for operational
9	Debts	Not allowed	Allowed
10	Financial reports	Government accounting standard (*Standar Akuntansi Pemerintah – SAP*)	Government accounting standard (*Standar Akuntansi Pemerintah – SAP*) and financial accounting standard (Standar Akuntansi Keuangan – SAK)
11	Financial reports	Audit by BPK	Audit by independent auditor
12	Long-term investment	Not allowed	Allowed
13	Purchasing	Based on presidential regulation	Has its own regulation
14	Cooperation	Major/governor (local government)	Head of working group

Source: Nugroho et al. (2014)

of KPH for their district. On the other side, running FMU needs independent financial support. State budget is limited, while to develop FMU as a full business entity will violate main objective of FMU. Proper format of FMU would be a quasi-government body like BLUD. Central government should pay attention more on strategic regulation for this PPK BLUD then.

According to the ministry of domestic affairs regulation No. 61/2007, BLUD is SKPD or nonprofit working unit under local government, which is established to support and provide services for the respective areas. BLUD has flexibility in budget management compared to conventional working units (Table 9.1). Legal procedure for retribution mechanism of BLUD is also rather less complicated compared to conventional working units (Regulation No. 28/2009). Retribution mechanism usually should be supported by local regulation – approved by DPRD (local parliament) but for BLUD only need major or governor decree.

In terms of giving more authority to FMU, based on inputs from local stakeholders, FMU which currently is only authorized to manage state forest area should also be given authority to manage non-forest area (CER Indonesia and CCAP 2010). By giving this authority, FMU can assist in managing REDD activities both within and outside forest areas. FMU should take the form of BLUD (*Badan Layanan Umum Daerah* – Local Service Unit). Having effective, strong, and independent FMU might be a key factor to the success of implementation of sustainable forest management.

9.4.2 Forest Certification System

Degradation of production forest is mainly triggered by the higher demand of wood for industry. Supply of timber from natural forests is not enough to meet the capacity of timber industry, and this leads to the increase of illegal logging activities. It is estimated that an additional supply of timber from illegal logging may be equal to that from the legal logging. The highest logging activities occurred in production forests (60 %) and then in the protected forest (30 %) and forest conservation (10 %). The level of illegal logging is estimated to be very high in the non-concession forest area of production forests (*Tim Pokja Kementrian Kehutanan* 2010). In other Asian countries, fuel wood collection and charcoal production for meeting domestic and local demand are also drivers of forest degradation (Hosonuma et al. 2012), even if their impact is not as much as that of illegal logging.

To reduce trading of illegal logs and to push application of sustainable forest management practices, the Government of Indonesia has established Timber Legality Assurance System (TLAS) through the issuance by the Minister of Forestry of Regulation Number P.38/Menhut-II/2009 on Standard for Evaluating Performance of Implementation of Sustainable Production Forest Management (PK-PHPL) and Verification of Legality of Logs (SVLK). This regulation is followed by the issuance by the Directorate General of Production Forest Regulation Number P.06/VI-Set/2009 and P2/VI-Set/2010. In TLAS, the assessment and verification of the timber products were done by independent third party, i.e., Entity for Evaluation of Performance and Independent Verifier (LP and VI) accredited by National Accreditation Committee (KAN). Other independent third parties such as Civil Society Organization and NGOs do the monitoring, i.e., for accommodating complaint from communities to the results of works from the LP and VI. With such process, TLAS will meet the good governance principles (transparency, accountability, and participatory), credibility (do not include government institution), and representativeness.

PK-PHPL is mandatory for all permit holders in state forests (IUPHHK-HA, IPPHHK-HT, IUPHHK-RE, HKm, and HTR) and private forests (Hutan Rakyat or

HR), and SVLK is mandatory for all permit holders in state forests and private forests and also for all upstream and downstream wood industries (IUIPHHK). In principle, permit holders who already have certificate of PHPL will not require to have SVLK. Validity of the certificate is only for 3 years, and every year it is subject to surveillance. Up to January 2013, total forest areas that have been granted for IUPHHK-HA (logging concessions), IUPHHK-HT (Industrial Timber Plantation), and IUPHHK-RE (Restoration of Production Forest Ecosystem) were 20,899,673 ha, 10.106.540 ha, and 397,878 ha, respectively. community forest plantations (HTR and HKm) were 752,297 ha (MoFor 2014).

In addition to the mandatory certification, there are also some voluntary certifications of SFM using standard *Lembaga Ecolabelling Indonesia* (LEI), Forest Stewardship Council (FSC), and some others. However, the progress of the implementation of certification is quite slow. Since June 2011, the total number of companies who already have mandatory certification of SFM had only been 230 certificates covering a total area of about 19 million ha and for voluntary certification had only been 25 certificates (Table 9.2). A number of factors that need attention for accelerating the achievement of SFM are (Bahruni 2011):

1. Governance and regulations which promote forest good behavior and reduce inefficiency of bureaucracy, encourage professionalism in forest management, push high responsibility of forest management units in using their given rights and authorities, and implement improvement program in organization capacity and forest management skill including resolving land uncertainty issues (tenure and spatial layout)
2. Provision of incentive and disincentives for forest management units with good performance and bad performance (SFM and non-SFM units) and allowing non-SFM units to improve their performance by planning and conducting concrete actions within clear timeline to meet SFM
3. Development of carbon accounting system to evaluate the performance of forest management units in minimizing forest degradation

Table 9.2 Number of companies who already have certification of SFM

		Mandatory certificates (up to June 2011)[2]		Voluntary certificates (up to June 2011)[3]	
Category	Total concession area (ha)[1]	Number	Area (ha)	Number	Area (ha)
IUPHHK-HA	22,710,256	140	14,225,443	5	834,452
Very good-good	*na*	*31*	*3,449,955*	*na*	*na*
Average	*na*	*35*	*3,307,789*	*na*	*na*
Poor or expire	*na*	*74*	*7,467,699*	*na*	*na*
IUPHHK-HT	9,963,770	90	4,914,301	3	544,705
Good	*na*	*19*	*2,499,280*	*na*	*na*
Expire	*na*	*71*	*2,415,021*	*na*	*na*
HR	1,570,315	Na	na	17	242,931

To encourage the concession holders applying for the certification, the government needs to revisit the SFM performance indicators used by forest management units (FMUs) that have different nature of activities, i.e., between management of forest resources (IPHHK-HA) and management of forest ecosystem (IUPHHK-RE; Nugroho et al. 2011). Different from IPHHK-HA, holders of IUPHHK-RE will have no cash inflow for a number of years until forests are restored as the timber will be harvested after reaching the equilibrium of ecosystems (e.g., 35 years). Applying for the certification will increase the cost, while the IUPHHK-RE holders are burdened with the obligation to pay various fees as applied to IUPHHK-HA. It is understandable that none of IUPHHK-RE (restoration of ecosystem) holders apply for the mandatory certification. A number of studies have proven that applying SFM practices will ensure the sustainable wood production and reduce the degradation (see Box 9.1). The reduction of emission from forest degradation by applying SFM practices could reach 9.79 tCO_2 ha^{-1} $year^{-1}$.

On the other hand, to conserve forests particularly forested land in forest area that have been released for non-forest-based activities, the Government of Indonesia also plans to apply mandatory certification system for palm oil called Indonesian Sustainable Palm Oil (ISPO). With this policy, all palm oil plantation companies will be obliged to conserve High Conservation Value (HCV) areas in their concession and to apply good practices in reducing GHG emissions. This policy is expected also to reduce deforestation. The ISPO will be officially effective as of March 2012, and it is targeted that all oil palm plantation companies will obtain the ISPO certificates by 2014. ISPO is launched to speed up the implementation of sustainable palm oil. ISPO is the same as existing sustainable standard RSPO (*Roundtable on Sustainable Palm Oil*); the only difference is that ISPO is compulsory, while RSPO is voluntary. Companies that have been certified by RSPO can receive ISPO certification after fulfilling some additional criteria. The regulation of ISPO is defined in the Ministry of Agriculture Regulation No. 19/Permentan/OT.140/3/2011. ISPO is a response of the Government of Indonesia to meet increasing demand of market for sustainable and green products and participate in mitigating climate change.

The mandatory certification system may also be followed by other non-forest-based activities that may directly affect forest resources such as mining. It has been well known that Indonesian forests store mineral deposits underneath which are needed to develop the country. Rights to use the resources are granted by the government through the scheme of *pinjam pakai* or land leasing for certain period of time. Mining of the deposit starts by clear off not only woody biomass of the forest but also other biomasses stored underneath the soil. The activities produces high emission which will be difficult to restore them back as fertility of the soil will be gone. In many cases, most of forest areas left by the mining after the termination of its permit are under heavily degraded condition.

To ensure the implementation of sustainable management principles and community economic development in exploiting natural resources (including mining), it may be necessary to introduce policy for limiting GHG emission (emission cap) from certain forest industries/concessionaires. The cap could be determined based on the result of the environmental impact assessment (EIA). Under current regulation, all forest industries/concessionaires obliged to conduct EIA. With the introduction of this policy, each entity must include the assessment of GHG emission level from their activities given all mandatory environmental management activities are met. Once the level of GHG emission is defined and estimated, this level of emission will be treated as "emission cap" of these entities. Theoretically entities that implement their environmental management plan defined in the EIA, the level of emission, should be low. Thus, companies that release more than the allowable emissions (emission cap) shall offset the excess.

Box 9.1 Impact of Sustainable Forest Management (SFM) Practices on Carbon Stock Change at Logging Concession Companies (Source: Bahruni 2011)

Based on data collected by Bahruni (2011) from five concessions (three concessions with SFM certification and two with non-SFM certification), it is quite clear that implementation of SFM practices can reduce emission from forest degradation. In non-SFM concessions, the volume of wood extracted relative to the annual allowable cut decreased significantly from year to year indicating continuous degradation of the forest, while in SFM concession, it is relatively constant (Fig. 9.7). Rate of forest degradation in SFM concessions was found to be between 0.17 % and 0.37 % per year and non-SFM between 2.35 % and 2.61 % per year and this equivalent to CO_2 emission reduction of 9.76 tCO_2 ha^{-1} $year^{-1}$ (Table 9.3).

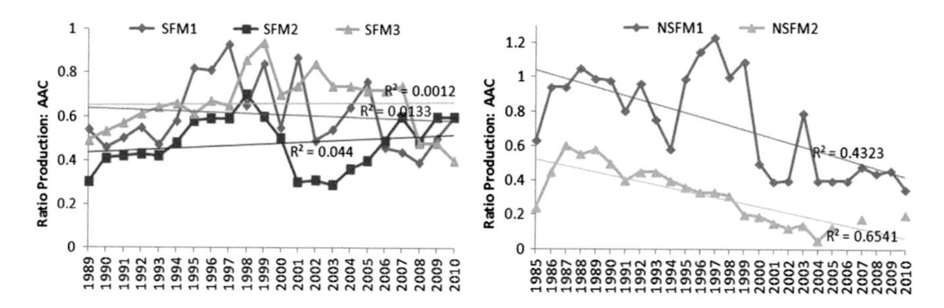

Fig. 9.7 Ratio between volume of wood extraction and annual allowable cut in SFN and non-SFM concessions

Table 9.3 Estimated CO2 emission reduction from forest degradation

Time period	The rate of degradation (%)		
	SFM	Non-SFM	Difference SFM and non-SFM
1992–2011	0.37	2.35	1.98
2000–2011	0.17	2.61	2.44
The benefit of SFM		1992–2011	2000–2011
The reduction of loss stand (m^3 ha^{-1} $year^{-1}$)		1.85	2.28
The reduction of emission (tC ha^{-1} $year^{-1}$)		2.16	2.66
The reduction of emission (tCO_2 ha^{-1} $year^{-1}$)		7.93	9.76

9.4.3 Reduction of Dependency on Natural Forests for Wood Supply and Sink Enhancement

In meeting wood demand, Indonesia already has begun to issue Timber Forest Product Utilization License (TPFUL) since the early 1970s, called as forest management right particularly for timber (forest concession or HPH). The highest number of concessions was in 1980 which is more than 500 units of concession with an area of 60 million hectares. After the enactment of Law No. 41 of 1999, forest concessions (HPH) were renamed as IUPHHK. Until now, the number of holders of IUPHHK for natural forests (HA) is declining to only about 256 units IUPHHK-HA. On the contrary IUPHHK for timber plantation (HT) increased from only a dozen units to 215 units by 2011, and community timber plantation (HTR) is also emerging with newly established plantation of about 0.63 million hectares involving more of 63 000 heads of households (HH).

HTI management unit is currently growing rapidly with total area more than 9.4 million hectares and targeted to grow to about 15.9 million hectares by 2030 (RKTN; Kemenhut 2011). Nevertheless community forest management (CFM) does not show significant development even though the Ministry of Forestry has set up high target (Table 9.4). So far IUPHHK-HTR that has been issued was only less than 100,000 hectares. Similarly both HKm and village forest also do not show significant improvement (Table 9.4). The schemes of HKm and HTR aim to revitalize the traditional wood-processing sector such as plywood and sawn-timber, in addition to increase the supply of raw materials for round-wood and paper and pulpwood industries. The program will enrich stock of carbon inside forest area by plantation activities done by smallholder farmers. It is expected by 2016 the plantations will meet its target to rehabilitate and improve productivity of degraded 5.4 million hectares of forest lands. Enrichment of forest carbon stock could be strengthened by investing the expansion of agroforestry system into the HKM and HTR schemes. On the other hand, private forest (Hutan Rakyat or HR) increased significantly only in Java, which is now reaching approximately 2.8 million hectares with production of about 6 million m^3 timbers per year. HR will continue to expand along with the proliferation of timber processing industry.

For increasing carbon sequestration, the Government of Indonesia has also implemented a number of programs for rehabilitating the degraded forest and

Table 9.4 Target, allocation, verification, and license issuance of community-based forestry up to 2010

Community-based forestry program	Target up to 2014 (Ha)	Allocation (Ha)	Verification (Ha)	License issuance by the Ministry of Forestry (Ha)	License issuance by the governor/head of district (Ha)
Community forestry (HKm)	2.000.000	400.000	203.573	80.181	30.485,55
Community forest plantation (HTR)	5.400.000	631.628			90.414,89
Forest village (HD)	500.000	179.187	144.730	13.351	10.310,00
Total	7.900.000	1.210.815		93.532	120.910,44

Source: Sub-Direktorat HKm, HD dan HTR Kemenhut (in Nugroho et al. 2011)

Table 9.5 Condition of production forest

Category	Production forest condition	Area (million ha)
1	Production forests with medium to very low level of degradation and now are still under management of concessionaires (IUPHHK-HA)	6.75
2	Production forests with medium level of degradation (no concessionaires operate in the area)	6.40
3	Production forest with medium to very high level of degradation (no concessionaires operate in the area)	14.15
4	Production forest with very high level of degradation (not meet forest definition anymore)	27.33
Total		55.62

Source: Based on MoFor (2014) and Bahruni (2011)

lands. At present due to the unsustainable practices of forest management, about 55.62 million hectares of production forest have been degraded (MoFor 2014). The level of degradation can be seen in Table 9.5. Production forests under categories 2 and 3 are allocated for restoration of production forest ecosystem. Up to 2013, total area of degraded production forests that have been granted with IUPPHK-RE was only 397,878 ha. To increase the interest of private sector to invest in the restoration of production forest ecosystem (IUPHHK-RE), the government may need to revisit its policy and regulations as RE activity has different nature of activities with IPHHK-HA. An incentive system should also be introduced.

As previously mentioned, the holders of IUPHHK-RE may not have cash inflow for a number of years until forests are restored as the timber will be harvested after reaching the equilibrium of ecosystems (e.g., 35 years). On the other hand, before the business permit is issued, they are burdened with the obligation to pay many fees as applied to IUPHHK-HA. In most cases, the holders of IUPHHK-RE can survive as they received grants from foreign donors who request for preservation of the forest ecosystem. Nugroho et al. (2011) recommended restructuring the regulations on forest ecosystem restoration by involving the managers of ecosystem restoration, government, and society. First is that ecosystem restoration business is not profit-oriented business so that the treatments should be different from IUPHHK-HA. Second, the current regulations PP. 3/2007 jo PP. No.03/2008 and ministry regulations should be revised to incorporate fundamental substantial changes, particularly on rights and obligations of license holders. Third is introducing incentive system for holders of IUPHHK-RE as they actually carry out government obligation in restoring, conserving, and preserving forests that nearly have no beneficial products.

Policy to prioritize the use of degraded forest for establishment of timber plantation will enhance sink as carbon stock of timber plantation is much higher than the degraded land and forest. In addition, the government for many years has also implemented a program for rehabilitating lands in forest area (*program reforestasi*) and non-forest area (*program penghijauan*). In the last 10 years, the Government of Indonesia has accelerated this program through GERHAN (*Gerakan Rehabilitasi Lahan dan Hutan*). In the period of between 2003 and 2008, total areas planted through GERHAN reached 1,767,559 ha or equivalent to about 300 thousand hectares per year or almost double than those implemented before this period. In the National Forestry Plan (RKTN; Kemenhut 2011), it is estimated that total degraded land in forest area that needs to be rehabilitated until 2030 is about 11.6 million ha. Therefore, rehabilitation of degraded land will be accelerated. Annually, it is targeted that at least 580 thousand hectares of degraded land is planted for rehabilitation.

Based on past experience in the implementation of the land rehabilitation program, it was found that the level of success of this program is still low due to lack of maintenance system (see Box 9.2) and no responsible management unit exists to maintain the planted trees. Without changing strategy in the implementation of GERHAN, the target being defined in the RKTN will not be achieved. For future program, the targeted area for GERHAN should be implemented in area where the FMU already exists, and whenever possible, its implementation should be integrated with CBFM program.

9.4.4 Reduction of Pressure on Natural Forest by Optimizing Land Use and Improving Land Productivity and Community Livelihood

In many regions, conversion of forest is mainly for agriculture activities either by community or by company. Community normally encroached to forest area illegally for planting annual crops or plantation, while company converted the forest to agriculture plantation after having permit. The encroachment occurred in all forest function but mainly in production forests. Therefore, many of forest areas are not covered by forest. On the other hand, the Ministry of Forestry releases conversion forest to local government to be used for non-forest-based activities where part of the area is still covered by forest, while the other part is already deforested and used by community. This condition often creates conflict between the community and company when local government issued permit to a company to use the land for plantation. Local government normally leaves the problem to company to solve, and this creates high social cost for the company. When this problem is not properly handled by companies, community will find new land and do encroachment again. In other case, communities expand their agriculture land through encroachment when their demand for land increases as the number of family increases. Looking at this condition, pressure on natural forest will continue if integrated efforts across related ministries and local governments are not in place.

Box 9.2 Survival Rate of Trees Under GERHAN Program

Based on assessment conducted by an independent consultant, PT Equality Indonesia on GERHAN Program implemented in 2006/2007 at West Java Province, it was found that the planted trees that can survive and form forest stand were only 20 % even the total area planted over 80 % of the target. On average based on evaluation in 13 districts in West Java Province, realization of GERHAN program reached 84 %, but the ones that survive were only about 53 % (note: *based on regulation from the Ministry of Forestry, the GERHAN program is considered to be successful if the survival rate over 56 %, without considering the condition of the trees*). Further evaluation indicated that of the 53 %, the survived trees with healthy condition were only 42 % (Fig. 9.8). Based on this condition, number of trees that can survive until forest stand on average will be about 18 % (0.84*0.53*0.42).

(continued)

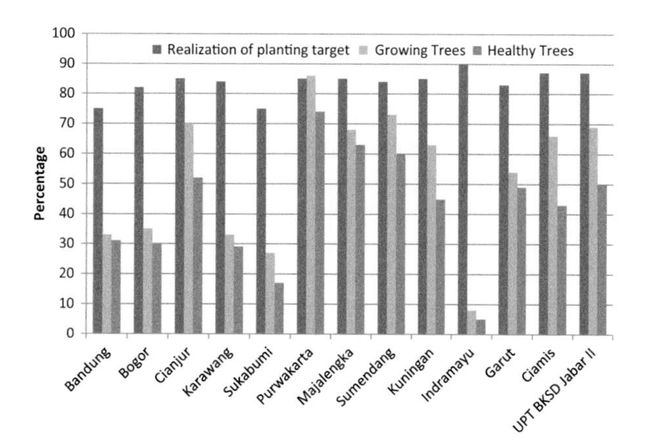

Fig. 9.8 Percentage of realization of planting area, survive trees, and healthy trees (Analyzed from data of PT. Equality Indonesia 2007)

Box 9.2 (continued)

Based on observation, implementation of GERHAN program in Java islands was relatively better than those outside Java. Considering these findings, it can be estimated that the level of success of GERHAN program may be around 20 %. If there is no change in the implementation system of the GERHAN program, with average planting rate of about 300 thousand hectares per year, GERHAN areas which are able to form forest stand will be only 60 thousand hectare.

Policies and potential programs that have been discussed and proposed by stakeholders in reducing threat on natural forests and deforestation include the following:

1. Enforcement of plantation companies to engage community in their plantation as plasma farmers. Regulation on this is already available, i.e., Ministry of Agriculture Regulation No. 26/Permentan/OT.140/2/2007 about Guidance on Permit for Agriculture Plantation. In this regulation every plantation company is obliged to establish plasma plantation at least 20 % of the total plantation area. However, many companies have not met this obligation. Following the implementation of mandatory certification system for plantations such as ISPO for palm oil, all companies are very likely to meet their obligation. In the case, where a company has already used all its land for plantation, the company will need to find land outside their plantation. If agriculture plantation commodities are allowed to be planted in forest area, this can be nicely integrated with community-based forest management (CBFM) program such as community timber plantation (HTR), community forest (HKm), and hutan desa (village forest). At present, one of agriculture plantation commodity allowed to be planted in forest area is rubber tree, while palm oil is still not allowed. In South Sumatra, HTR program has been implemented in reforesting production forest area using rubber tree.
2. Improvement of crop productivity of small holder farmers. Most of communities that occupy forest area for agriculture activities are poor farmers and have little knowledge in good agriculture practices. For example, based on discussion with farmers who occupy Kerinci Seblat National Park (KSNP) in South Sumatra, it was stated that community tended to expand their agriculture lands to meet food demand and income of their family as their family is growing. By increasing crop productivity, the demand for land is expected to decrease (see Box 9.3). Creation of other alternative income for this community as well as their institutional capacity can increase the effectiveness of this program in reducing pressure on the forest. Development of synergy or integration of community empowerment programs from various sector and private (CSR) would be needed to enhance the effectiveness of this program.
3. Optimization of the use of non-forested land for agriculture activities by changing forest function. As shown in Figs. 9.1 and 9.2, more than 10 Mha of land in conversion forest is forested land, while about 20 Mha land in production forests is

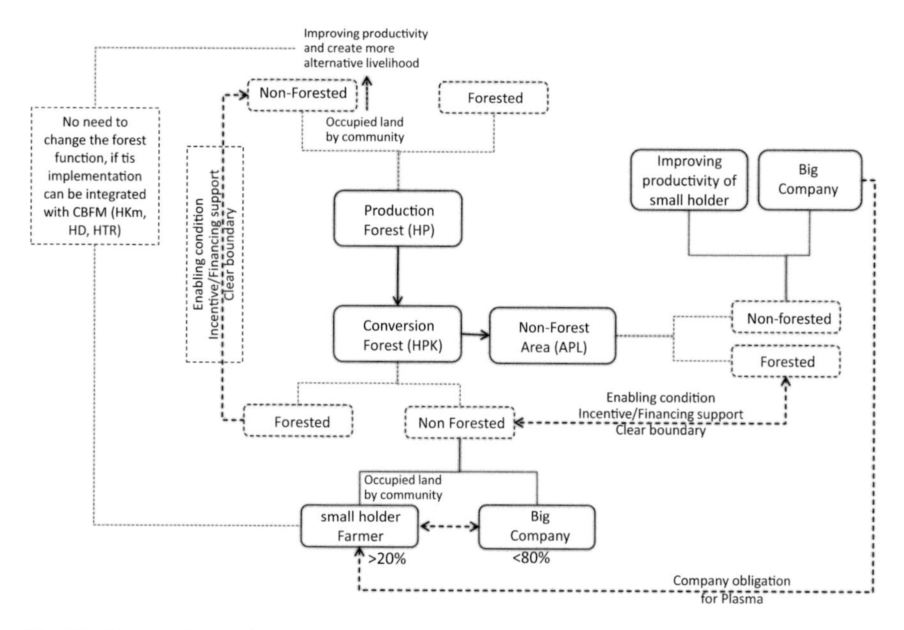

Fig. 9.9 Process for implementation of policy and program for reducing threat on natural forest and rate of deforestation (Modified from Boer et al. 2012)

non-forested land. In non-forest area, almost 7 Mha is forested land. Changing functions of forested conversion forest to production forest, and non-forested production forest to conversion forest which later can be released for non-forest-based activities (mainly for agriculture plantation) or swapping forested land in APL with non-forested conversion forest, would reduce future deforestation. Based on discussion with the staff of Planning Agency at Central Kalimantan Province, swapping forested land in non-forest area with non-forested production forest will be very difficult. It is suggested that before this land swap policy is applied, the status of non-forested production should be changed first to conversion forest. The Joint Minister Decree may be needed to implement this policy (Ministry of Forestry, Ministry of Internal Affairs, and National Land Agency). New direction on the utilization of forest area has been issued by the Ministry of Forestry in the RKTN (National Forestry Plan for 2011–2030), and this may need to be revisited if the policy is to be implemented. This land swap policy will also be potential to be integrated with mandatory certification and CFM programs. Obligation for agriculture plantation companies to develop plasma plantation with community with minimum area of about 20 % of the total plantation may need additional lands. If agriculture plantation commodities are allowed to be planted in forest area, there is no need to change the status of forest function, and this program can be integrated with the CFM programs. Collaboration between the Ministry of Forestry and Ministry of Agriculture is required to facilitate this program.

Figure 9.9 summarized the process of implementing policy and programs for reducing threat on natural forest and deforestation.

9.5 Financing and Incentive Policies for Supporting the Implementation of SFM and REDD+

To support the implementation of the above four key policies and actions, there are a number of financing and incentive policies that may need to be considered. These include (1) financing policies for the acceleration of FMU establishment, (2) incentive policies for the certification system, (3) financing and incentive policy for accelerating the establishment of timber plantation on degraded land and CFM for sink enhancement, and (4) incentive and financing policies for conserving forest carbon and land swap.

9.5.1 Financial Policy for Development of FMUs

As discussed above, the total number of FMUs that need to be established in Indonesia is about 600 units. Following target defined in the National Action Plan for reducing GHG emission (Appendix President Regulation 61/2011) within 5 years (2010–2014), the total number of FMUs that will be established is 120 units. With the total number of 600 FMUs, the time required to complete the establishment of FMU all over Indonesia would be 25 years. It is long process, with assumption that rate of deforestation in the future under the absence of FMUs follows historical rate; within the next 25 years, about 25 Mha of forest may be lost. Following Bappenas' assumption that the cost for establishing an FMU with self-funded capacity is 40 billion IDR (5 years), total cost required for the 600 units will be about 24 trillion IDR or 2.7 billion USD. Considering that this program will be a key for the success of REDD+, acceleration of FMUs establishment is necessary. Clear Roadmap on the Establishment of the all FMUs should be developed with secure budget. The Government of Indonesia may negotiate with donor countries to use debt-for-nature swap (DNS) scheme to secure budget to support the establishment of the FMU.

Box 9.3 Reducing Pressure on Kerinci Seblat National Park (Source: CER Indonesia and CCAP 2011)

Kerinci Seblat National Park (KSNP) is a part of the Bukit Barisan Mountain Range, stretching north to south along Sumatra Island. The park's location makes KSNP one of the richest conservation areas in terms of biodiversity. However, KSNP is under great threat of deforestation and forest degradation. A number of square kilometers of forest have been lost annually in the national park, severely reducing the natural environment for animals and other forest-dwelling life. The main drivers of deforestation and forest

(continued)

Fig. 9.10 (**a**) Slash and burn activity in KSNP, (**b**): agricultural land inside KSNP

Box 9.3 (continued)

degradation in KSNP are encroachment by the community for agricultural activities, illegal logging, and fires (Fig. 9.10).

Most of the villagers surrounding KSNP are involved in agricultural activities such as rubber and coffee production. Each household has 1–25 ha of land for agriculture; although illegal, some of this is done inside the KSNP area. Villagers enter the park because they need a large amount of land for agriculture. The productivity of coffee is very low, i.e., only 0.4 ton/ha or about one twentieth of normal yield (6–10 ton/ha). By increasing productivity of the crop just up to 4 t/ha will reduce the demand for land by ten times from the usual one. At least there are four programs that can be implemented for improving communities' agriculture practices, namely, (1) improving seed quality as in the usual practices communities get seed from forest or from their garden, (2) improving maintenance and inputs as in the usual practices farmers do not use fertilizer and there are no regular weeding and spraying, (3) improving timing for harvesting to improve quality of coffee as in usual practices farmers tend to harvest the coffee before it gets mature, and (4) improving post-harvest management.

Enhancing capacity of farmers for improving agriculture practices could increase productivity and their income and thereby reduce the demand for land. This can be expected to reduce deforestation in KSNP. Strong assistance for the community will be essential to maintain KSNP.

The roadmap for the establishment of FMU may include at least the following aspects: (1) development of criteria and indicator for prioritizing forest area for FMU's establishment, (2) strategy on FMU institutional capacity building, (3) development of strategic work plan of the FMU, and (4) monitoring and evaluation system. The first aspect is very important to develop as level of risk and problems vary across regions. The availability of criteria and indicator will help the government in putting priority where FMU should be first established and ensure the presence of FMU will have significant impact on the improvement of performance in forest management or keep good forest management system to continue. The second aspect refers to steps of actions that would be implemented in developing capacity of the FMU organization. The third aspect refers to readiness of the FMU to carry out its role and function, and the fourth aspect refers to development of system to monitor and evaluate the performance of the FMU which will be needed for the development of improvement plan of the FMU. Kartodihardjo et al. (2011) proposed at least eight criteria for evaluating the FMU development performance, namely, (1) area stability, (2) forest use planning, (3) management plan, (4) organizational capacity, (5) inter-strata relations within government and regulations, (6) investment mechanism, (7) availability of access and community rights, and (8) forestry dispute settlement mechanism. In each typology indicators need to be developed for these criteria.

In terms of FMU organization capacity, capacity development should enable the FMU (1) to promote forestry professionalism and be able to perform management that can produce economic value from forest utilization that is balanced with the conservation, protection, and social functions of the forest; (2) to develop investments and provide work opportunities; (3) to prepare spatial-based planning and monitoring/evaluation; (4) to protect forest interests (including the public interest in the forest); (5) to respond to the range of local, national, and global forest management impacts (e.g., the forest's role in mitigating global climate change); and (6) to adjust to local conditions/typology as well as strategic environmental changes affecting forest management (Kartodihardjo et al. 2011).

9.5.2 *Incentive System for Certification*

As discussed above, the Government of Indonesia has issued a number of mandatory certification systems. These mandatory certification systems as mentioned previously are applied for all forest management/business entities (from large to small scale), namely, IUPHHK-HA, IUPHHK-HT, and IUPHHK-RE, and community forest management (CFM),[3] namely, HTR and HKm (with permit utilization) and/or village forest/Adat forest (with management rights) and Hutan Rakyat (private forest, forest management on an owned land) as well as wood industries. For community-based forest management entities, obligation for doing certification may add burden as this will increase cost of production. On the other hand, some also argued the effectiveness of applying mandatory forest certification system, such as SFM/SVLK, in reducing illegal logging may also not be significant as the

[3] Community forest management (CFM) combines two things: a type of resource (forest) and a group of owner/manager (community). The term CFM broadly refers to various forms: Participatory Forest Management (PFM), Joint Forest Management (JFM), joint forest management (forest co-management), and Community-Based Forest Management (CBFM).

certified company only able to manage the illegal activities within its company site, while market for illegal wood still exists.

Applying same rules for IUPHHK-RE (ecosystem restoration) as applied to other wood business forest activities in certification process may also be counter-productive. In the IUPHHK-RE, forest management units (concessionaires) are not allowed to do wood logging until forest reaches equilibrium conditions (may take time for about 35 years). Thus, in the short term, there will be noncash inflow to the concessionaires. While at present, treatments in term of fee and procedure for getting the permit (IUPHHK-RE) are similar to IUPHHK-HA (HPH) and IUPHHK-HT (timber plantation) as well as obligation for having certification. Without changing this policy, interest of private to do investment for production forest ecosystem restoration will be very low. Based on data from Purnama and Daryanto (2006), more than 10 million ha of production forest is suitable for IUPHHK-RE, while until now the total area of degraded production forests granted with IUPPHK-RE was only 185,005 ha.

Another mandatory certification system for agriculture plantation such ISPO which will oblige plantation companies to develop plasma farmer with minimum area of 20 % of the total area of the plantation will also face dilemma. For new plantation, it may not be difficult to establish such plasma; however, for old plantation, this will be difficult as all their plantation areas are already planted. The only alternative ways is to find additional lands to be used for plasma. This will be very costly if no support mechanism from government.

Considering the above dilemma, incentive system for certification may need to be expanded. Some of potential incentive policies in supporting the mandatory certification system may include the following:

1. Expanding type of incentive for small business entity in getting certification. In the context of SVLK, program for increasing awareness of community on the importance of using certified wood product for saving environment should be progressively implemented. In reality, many wood consumers in developing nations do not care too much on this issue; the consumers are more interested in buying cheaper products. Based on discussion with stakeholder in East Java, price of illegal wood could be half of that the legal ones, so that wood products produced from these will be much cheaper. At present, the government has provided support for small holder company via government budget (APBN) to cover the cost for certification. This subsidy is still not enough as the cost for producing one unit product from certified timber is still higher than the one used illegal ones. In this regard, the incentive[4] for small holder may need to increase so that the price of certified wood product can compete with the non-certified one. At the same time, the awareness-raising programs for community for consuming certified wood products have to be promoted. The subsidy can be

[4] Incentive could also be given in form of direct inputs subsidy.

gradually reduced when domestic market for certified wood products increases. This type of policy could be also negotiated for debt-for-nature swap program.
2. Providing subsidy for business entities focusing on ecosystem restoration in having the mandatory certification.
3. Providing incentive for plantation companies in getting lands for plasma farmers as support for the company in meeting certification obligations. Implementation of this policy could be integrated with CFM programs.

9.5.3 Incentive and Financial Policy for Accelerating the Establishment of Timber Plantation on Degraded Land and CFM for Sink Enhancement

Many of degraded lands in forest area are claimed by community. When permit for using the land has been granted to an entity, conflict on the land normally emerges between the entity and the communities. For this reason, private entities prefer to use forested land in forest area for timber plantation or peatland as these areas normally have no or less conflict (no community claim on the land). Ideally, the government should issue permit on safe and conflict-free (clean and clear) forest areas. However, in most cases this is not the case, and the permit holders have to solve this land conflict problem. Level of conflict varies between regions, and social cost that has to be covered by the permit holder in the establishment of plantation will also vary. The high social costs prevent the permit holders to establish plantation. In this regard, the government needs to create incentive system for permit holders in handling this land conflict problem, and the types of the incentive may be varied depending on level of conflicts. The incentive could be in the form of reducing or exemption of administration/retribution fees for certain period of time depending on level of conflict. With this incentive policy, establishment of timber plantation in degraded land can be accelerated, and the dependency on natural forest for supplying wood will also reduce.

In managing the land conflict issue, the MoF also implements CFM program. The program gives access and right to communities to use the forest area or formalize/legalize the use of the land by the community. The communities have to apply for getting the permits (HTR, HKm, village, and Adat forests). However, the process for getting the permits is too complex for communities, and it is also a lengthy process. Without any assistance from their partners, communities are mostly unable to have the permits. Financial support from the government to communities in implementing the CFM is also available via BLU-P3H (General Services Agency). The amount of funding available for supporting the CFM is also huge, i.e., over a billion of USD. However, the absorption of fund is also very low, less than 1 %. Simplifying the process of getting permit and accessing fund from the BLU-P3H will also be crucial for accelerating the implementation of the CFM. As mentioned above, up to now the realization of the CFM program is far from target (see Table 9.4).

Acceleration of the ecosystem restoration program which will have significant contribution to sink enhancement also needs incentive from the government. Incentive in the form of reducing administration/retribution fees for certain period or exemption from some of administration/retribution fees is recommended.

9.5.4 Incentive and Financial Policies for Conserving Forest Carbon and Land Swap

Implementation of land swap policies and exchange of forest functions in order to avoid deforestation (conserving carbon stock in forest) will need incentive and financial policy supports. Nurrochmat (2011) proposed a number of incentive and financial policy for supporting local government in implementing the policies. These include:

1. Financial policy on special allocation fund (Dana Alokasi Kusus, DAK) for conservation. This policy is an incentive from the national government to local government that commits to conserve forest for environmental services. Special allocation fund given to the conservation region should compensate the benefit loss coming from natural resource extraction or forest land conversion (conversion value). The Ministry of Finance plans to accommodate this in revision of Act No. 33/2004 (Ministry of Finance 2011).
2. Revision of fiscal balance law to enforcing "liability rule." The present fiscal balance law regulates the benefit sharing of natural resource extraction between national and local governments, as well as among local governments. The magnitude of sharing depends on the magnitude income that comes from the extraction of natural resources. In this case, the higher the number of the natural resources extracted by certain region, the bigger the benefit sharing received by the region. Revision of the existing fiscal balance law to be a more green fiscal balance is needed to avoid overexploitation and further destruction of natural resources in the regions due to short-term economic interest. A green fiscal balance shall give a proportional attention both in the reward side and in the punishment side to ensure the sustainability of nature resource management.

From the above discussion, it is quite clear that the issue of forest boundary (safe and conflict-free forest areas) and policy on the issuance of permit on the use forest area are two factors that will contribute to the achievement of SFM and REDD+ implementation. Development of boundaries between non-forest and forest areas needs acceleration. In regard with the forest boundary issues, Kemenhut (2011) reported that up to 2010 length of boundary between forest and non-forest areas and between forest functions reach 281,873 km covering area of about 14,238,516 Ha or about 10 % of total forest area of Indonesia. This condition is considered as one of the important factors causing conflict of land right and access in all provinces. At present there is about 22.5–24.4 Mha of forest area in conflict, and a number of villages within forest area reach 19,420 villages (Dephut and BPS 2009 in Kartodihardjo et al. 2011).

The cost of developing forest boundary is quite expensive. Following the regulation from the MoF, cost for changing forest functions that include developing forest boundary is 3.4 billion IDR per 12,000 ha. To reduce the cost, the process of the development of the forest boundary could be integrated with the development of FMU and conducted through participatory mapping process. In line with recommendation from Kartodihardjo et al. (2011), in addressing this boundary issue in connection with FMU establishment, there are several strategic directives that should be adopted depending on conditions in the FMU. These directives include:

1. Localization of all areas that have serious tenurial conflict into areas of noneffective production as a transitional policy and gradually building a collaboration to optimize achievement of sustainable forest management objectives.
2. Development of micro-spatial arrangements together with the community in order to reach mutual agreement with the community on the utilization norms for each spatial function.
3. Recommendation of legal settlement through the mechanism of revising the spatial arrangements in areas with serious tenurial conflict that is unlikely to be retained as forest areas.
4. Accommodation of community access to forest resources by rearranging the norms for utilizing such resources in accordance with sustainability principles.
5. Development of a mechanism for recognizing community management rights in areas of serious/minor tenurial conflict in the context of sustainable forest management. This mechanism serves as the basis for FMU managers to prepare licensing recommendations for communities.
6. Engagement of law enforcement for all issues relating to illegal activities.

Another important key factor for achieving SFM is availability and accessibility of funds for supporting SFM practices, particularly for engagement of communities in CFM. With the current system, the available fund to support CFM managed by the BLU-P3H as discussed above is not easily accessed by community due to the administration procedure. Policy allowing for transferring the funds to a financing system relatively easy to be accessed by community is required. Two types of financing systems that can be generated at regional level and may meet this need are "blending financing" and "hybrid micro-financing systems" (CER Indonesia and CCAP 2010). Blending financing system is a financing system that synergizes all financial sources such as CSR funding, government funding such as state budget (APBN), and local government budget (APBD) funds, banking, and international funding. This system can help leverage private funding and supports regional development by supporting community activities in urban agriculture and agroforestry including building human resource capacity through assistance and training activities.

Unlike the blending financing model, the hybrid micro-financing system will utilize more government funds than private funds. Funding to support CFM (HTR, HKm, HD/HAd), which is currently managed by BLU-P3H, would be part of this

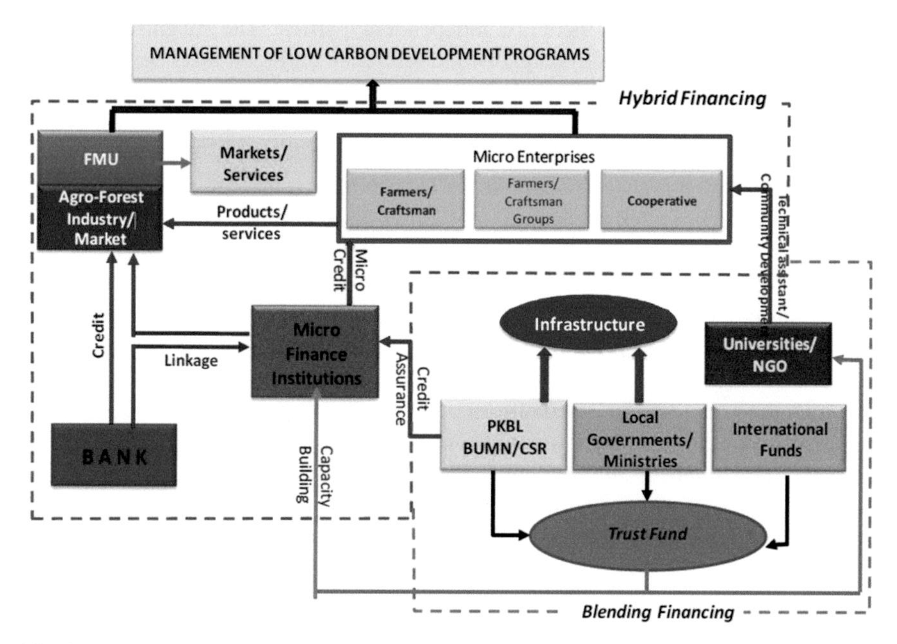

Fig. 9.11 Structure of financing systems to support low-carbon development (SFM and REDD+) (CER Indonesia and CCAP 2010)

financing system. In this system, government funds can be accessed by communities as capital fund assistance in the form of business credit. This system will require credit assurance institutions (LPKD – Local Credit Assurance Institution). The LPKD will provide government guarantees to banks so that if farmers are unable to pay on time, the LPKD will cover the credit and the farmers would pay later following rules as stated in Presidential Regulation No. 2/2008. This credit assurance institution has been developed in a few regions. The presence of this credit assurance institution is expected to support small- to medium-scale community business investments. Figure 9.11 presents the structure of the financing models and the connection with CSR and FMU. In the context of REDD+, both blending financing and hybrid micro-financing systems should provide positive incentives (low interest, tax deduction, concessional investment, etc.) for communities who propose activities that result in emission reductions from deforestation and degradation, conserving forest carbon, sustainable forest management practices, and sink enhancement.

Introduction of the incentive systems should not reduce the income of the government as the emission reduction which resulted from these policies will result in carbon payment. As decided by the Conference of the Parties (COP), payment from the implementation of REDD+ activities will be performance-based payment. This means that the party who decided to join the REDD+ scheme will be eligible to get the payment after the achievement in reducing the emission is measured, reported, and verified by the third independent parties. The magnitude of the emission reduction is measured against the reference emission level being used.

The study of Schmitz et al. (2014) indicated that by increasing investment in forest sector for facilitating change in technology (TC) at rate of 1 % per year on top of the external investment, the forest destruction might decrease. The hypothesis of this scenario is that higher investments in TC can reduce the rate of forest destruction without any forest protection (e.g., investing in agricultural productivity reduces pressure on tropical forests without the necessity of direct protection; see section 9.6.5). Their study suggested that in the Pacific Asia (most of forest in Indonesia) without any significant change in forest protection program from Business as Usual (called as reference scenario), by 2050 the loss of forest cover might reach 43 % of that of 2010. If this figure is used for Indonesia, under the reference scenario, the remaining Indonesia forest cover by 2050 would be about 55.7 million ha or equivalent to total forest loss of about 43.4 million ha or about 1.08 million ha per year. This figure is slightly higher than the historical deforestation rate in the period 1990–2013, i.e., 0.822 million ha per year (see Sect. 9.3). With the increase of investment by 1 % per year from the top of the external investment for the TC, it is expected that the forest loss in Indonesia would be only 13.5 million ha or equivalent to rate of deforestation of 0.337 million ha per year. As discussed above, the implementation of the above innovative financing and incentive policies might improve land and forest management, and this may further increase the potential of reducing emission from REDD+ activities.

References

Avisar R, Werth D (2005) Global hydroclimatological teleconnections resulting from tropical deforestation. J Hydrometeorol 6:134–145

Bahruni (2011) Conduct study and analysis on economic incentive framework of SFM as important option for forest based climate change mitigation-to reduce emission from and by tropical forest. ITTO Project Report RED-PD 007/09 Rev.2 (F)

Bappenas (2010) Naskah Akademis Rencana Aksi Nasional Penurunan Emisi Gas Rumah Kaca (GRK) 2010–2020. Badan Perencanaan Pembangunan Nasional, Republik Indonesia, Jakarta

Boer R (2009) Reduction emission from deforestation and forest degradation and sustainable development in Indonesia. In: Habito CF, Kojima S (eds) Mainstreaming sustainable development policies in East Asia. ERIA research project report 2008 No. 6–2. http://www.eria.org/pdf/research/y2008/no6-2/Appendix3.pdf

Boer R (2012) Sustainable forest management, forest based carbon, carbon stock, co_2 sequestration and green product in order to reduce emission from deforestation and forest degradation. Technical Report: Indonesia's Ministry of Forestry International Tropical Timber Organization RED-PD 007/09 Rev. 2 (F): Enhancing Forest Carbon Stock to Reduce Emissions from Deforestation and Degradation through Sustainable Forest Management (SFM) Initiatives in Indonesia, Jakarta

Boer R, Nurrrochmat DR, Ardiansyah M, Purwawangsa H, Hariyadi, Ginting G (2012). Reducing agricultural expansion into forests in Central Kalimantan-Indonesia: analysis of implementation and financing gaps. Research report submitted to Prince of York. Bogor

Busch J, Ferretti-Gallon K, Engelmann J, Wright M, Austin KG, Stolle F, Turubanova S, Potapov PV, Margono B, Hansen MH, Baccini A (2015) Reductions in emissions from deforestation from Indonesia's moratorium on new oil palm, timber, and logging concessions. PNAS 112:1328–1333

CER Indonesia and CCAP (2010) Establishing integrated forest policies to reduce greenhouse gas emissions from deforestation and forest degradation at the District Level. Phase 1: the District of Musi Rawas, South Sumatra. Carbon Environmental Research Indonesia (CER Indonesia) and Centre for Clean Air Policy (CCAP), Bogor

CER Indonesia and CCAP (2011). Establishing Integrated Forest Policies to Reduce Greenhouse Gas Emissions from Deforestation and Forest Degradation at the District Level. Phase 2: The District of Musi Rawas, South Sumatra. Carbon Environmental Research Indonesia (CER Indonesia) and Centre for Clean Air Policy (CCAP), Bogor

Dale VH (1997) The relationship between land-use change and climate change. Ecol Appl 7:753–769

Dickinson R, Henderson-Sellers A (1988) Modelling tropical deforestation: a study of GCM land-surface parameterizations. Quart J R Meteor Soc 114:439–462

Directorate of Forest Resource Inventory and Monitoring (2015) Data dan Informasi Geospasial Dasar dan Tematik Kehutanan Terkini Tingkat Nasional. Presentation for Workshop on "Working Preparation for 2015", Auditorium Utama Manggala Wanabakti, 2 Maret 2015, Jakarta

Ditjen BUK (2011) Data Release Ditjen BUK Triwulan 1 Tahun 2011. Direktorat Jendral Bina Usaha Kehutanan, Jakarta. www.dephut.go.id

Hergoualc'h K, Verchot LV (2014) Greenhouse gas emission factors for land use and land-use change in Southeast Asian peatlands. Mitig Adapt Strateg Glob Change 19:789–807

Hosonuma N, Herold M, De Sy V, De Fries RS, Brockhaus M, Verchot L, Angelsen A, Romijn E (2012) An assessment of deforestation and forest degradation drivers in developing countries. Environ Res Lett 7 (12pp). doi:0.1088/1748-9326/7/4/044009

Houghton RA, House JI, Pongratz J, van der Werf GR, De- Fries RS, Hansen MC, Le Quéré C, Ramankutty N (2012) Carbon emissions from land use and land-cover change. Biogeosciences 9:5125–5142. doi:10.5194/bg-9-5125-2012

Kartodihardjo H, Nugroho B, Putro HR (2011) Forest Management Unit Development (FMU): concept, legislation and implementation. Directorate General of Forestry Planning, Ministry of Forestry, Jakarta

Kemenhut (2011) Rencana Kehutanan Tingkat Nasional (RKTN) tahun 2011–2030. Kementrian Kehutanan Republik Indonesia. Jakarta

Le Quere C, Andres RJ, Boden T, Conway T, Houghton RA, House JI, Marland G, Peters GP, van der Werf GR, Ahlstrom A, Andrew RM, Bopp L, Canadell JG, Ciais P, Doney SC, Enright C, Friedlingstein P, Huntingford C, Jain AK, Jourdain C, Kato E, Keeling RF, Klein Goldewijk K, Levis S, Levy P, Lomas M, Poulter B, Raupach MR, Schwinger J, Sitch S, Stocker BD, Viovy N, Zaehle S, Zeng N (2013) The global carbon budget 1959–2011. Earth Syst Sci Data 5:165–185. doi:10.5194/essd-5-165-2013

Ministry of Finance (2011) Policy brief: instrument and financing mechanism on reducing GHG emission. Fiscal Policy Agency, Ministry of Finance, Jakarta

MoE (2003) National strategy study on CDM in forestry sector. Ministry of Environment, Jakarta

MoE (2010) Indonesian second national communication to UNFCCC. Ministry of Environment-Republic of Indonesia, Jakarta

MoFor (2014) Ministry of forestry statistics year 2013. Ministry of Forestry, Jakarta

Nugroho B, Sukardi D, Widyantoro B (2011) Study and analyze regulations concerning sustainable forest management, forest based carbon, carbon stock, CO2 sequestration and green product. ITTP Project Report RED-PD 007/09 Rev.2 (F)

Nugroho B, Ridwan M, Hendri, Kartikasari K, Boer R (2014) Development of FMU as one of the strategies for ensuring the sustainability of emission reduction commitment. Technical report for LAMA-I Project, ICRAF, Bogor

Nurrochmat DR (2011) Review infrastructure framework and mechanism related to SFM as Important option in reducing emission from deforestation and forest degradation. Report submitted to Indonesia Ministry of Forestry and International Tropical Timber Organization, Jakarta

PT. Equality Indonesia (2007) Laporan akhir Pekerjaan Penilaian Tanaman GN-RHL/Gerhan tahun 2006/2007 Wilayah Propinsi Jabar. Bogor

Rusolono T, Tiryana T (2011) Review of existing sustainable forest management (SFM)-based projects in Indonesia. ITTO Project Report RED-PD 007/09 Rev.2 (F)

Schmitz C, Kreidenweis U, Lotze-Campen H, Popp A, Krause M, Dietrich JP, Muller C (2014) Agricultural trade and tropical deforestation: interactions and related policy options. Reg Environ Change, 16pp. doi:10.1007/s10113-014-0700-2

Shukla J, Nobre C, Sellers PJ (1990) Amazon deforestation and climate change. Science 247:1322–1325

Tim Pokja Kementrian Kehutanan (2010) Laporan Kelompok Kerja Kebijakan Kehutanan. Kementrian Kehutanan, Jakarta

Chapter 10
Fostering Capacity Development for Asia's Leapfrog

Sirintornthep Towprayoon

Abstract *Capacity Development Is the Basis for Asia's Leapfrog*

Most major development paths of Asian countries are moving toward green growth. Under the constraints of the energy crisis and climate change impact, future Asian growth, while appearing to be the most significant in comparison with other regions, needs a good knowledge-based pathway to light up and pave the road to a low-carbon society.

Capacity development is the basic need and a urgent issue to be explored in Asia. It is one of the effective tools for Asia to leapfrog to a low-carbon society with the concern of unlocked carbon intensity development.

A Bullet Train Model

Development needs to be done on several levels from communities to the subnational and national levels. To leapfrog from the current situation, capacities need to be built at many levels through various mechanisms of networking, research forums, initiatives, training, etc., in order to bridge, transfer and transform the results from research to policy and to implementation. Policy makers with good understanding, as the head of the bullet train, will lead society in the right direction, while scientists and researchers are the engines to back up and accelerate this movement. Finally, practitioners in communities play key roles as the fuel, enhancing the movement toward green growth through their activities. It is, therefore, essential to have these three components for a compatible basis of knowledge and comprehension through capacity development.

Asian Countries Need Collaboration

Asian countries are different in nature but rich in culture and resources. Low-carbon activities are various and depend on internal factors and situations. There are many good practices and philosophies that can be shared among the countries. The experience of learning from each other facilitates accomplishments and reduce risks in implementation. Collaborative activities in capacity development help Asian countries move toward green growth in their own ways with their own uniqueness while seeing the same goal in the future together.

S. Towprayoon (✉)
Joint Graduate School of Energy and Environment, Center of Excellence on Energy Technology and Environment, King Mongkut's University of Technology Thonburi, Bangkok, Thailand
e-mail: sirin.jgsee@gmail.com

S. Nishioka (ed.), *Enabling Asia to Stabilise the Climate*,
DOI 10.1007/978-981-287-826-7_10

Keywords Capacity building • Collaboration • Asian leapfrog • Research to policy • Low carbon • Knowledge-sharing platform

Key Message to Policy Makers

- Capacity development is a basic need and an urgent issue for Asia's leapfrog.
- This can be done through knowledge transfer, research collaboration and joint education programs among Asian countries.
- A full loop of knowledge transfer from research to policy and to implementation is the key to success for capacity building in Asia.

10.1 Capacity Development Is Important in Asia—A Tool for Leapfrog

This chapter will explain the need to have capacity development (CD) in Asia, which is a so-called tool for leapfrogging to a low-carbon society. Comparison of emissions from the past and future projections will be drafted. The nature of Asian countries in terms of population and competition, resource utilization, understanding of the people on the ground, and some philosophies of implementation in specific countries will be illustrated. This will lead to the conclusion of capacity building for leapfrogging to a low-carbon society. Entering into a low-carbon society for developing countries is difficult. While there is the potential for Asia to lead climate change abatement, implementation is the key and this cannot be achieved if there is no capacity building at all levels.

10.1.1 The Power of Asia

The merging of 10 ASEAN countries (Brunei Darussalam, Cambodia, Indonesia, Lao PDR, Malaysia, Myanmar, Philippines, Singapore, Thailand, Vietnam) into the ASEAN Economic Community (AEC) from 2015 onward makes ASEAN become more important in the Asian region. These ten countries of the AEC contribute 9 % (585 million people) of the world's population while their GDP contributes 3 % (1275 billion USD) of the global GDP. In addition, the ASEAN +3, with the coverage of China, Japan and South Korea, increases the share of the world's population by 31 % (2068 million people) with 18 % (9.901 billion USD) of the global GDP. The immense contribution with a high impact on economics can be

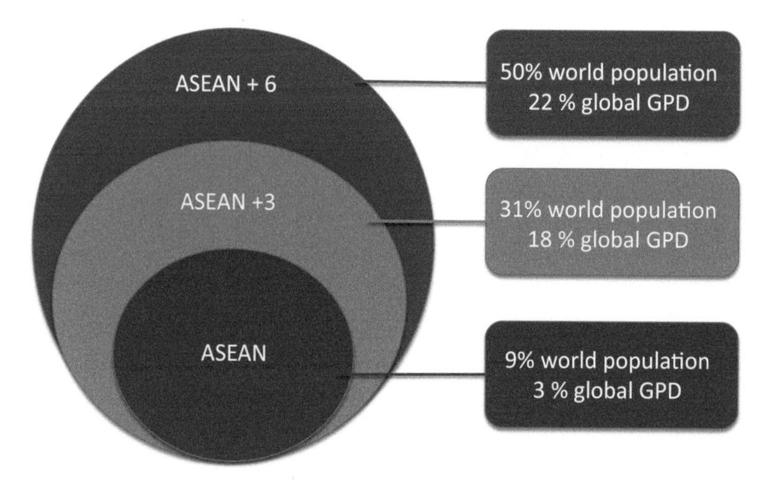

Fig. 10.1 Shares of population and GDP in the global situation

seen with the ASEAN +6 (Australia, China, India, Japan, New Zealand, South Korea) where the share of the world population is 50 % (3284 million people) and its contribution covers 22 % of the global GDP. Figure 10.1 indicates that the path of movement into the future of ASEAN and the Asian countries will have an impact on the world's development.

10.1.2 The Rise of ASEAN

The nature of the ASEAN countries varies, particularly their economic conditions. Regarding the classification of income in ASEAN countries, the composition of income classification is shown in Fig. 10.2. Four categories of income have been identified. Singapore and Brunei are identified as affluent countries while Malaysia and Thailand belong to the middle class. Three countries are in the process of transitioning to the middle class and the remaining three countries fall within the low-income range. The interesting aspect is that the ASEAN middle-income class is more than 25 % of the ASEAN population and in the year 2030 it is anticipated that the middle-income class segment in Indonesia will include more than 50 million people. While the share of the middle-income class increases, development of countries to move toward a middle-income trap has been raised in some countries like Thailand and Malaysia. How the rise of ASEAN can gear its direction toward a low-carbon society is challenging.

Fig. 10.2 Classification of incomes in ASEAN countries

10.1.3 Regional Development

While moving to middle incomes but aiming to avoid a middle-income trap, countries' development is still based on energy consumption. High energy use has been found to be related to the human development index as seen in Fig. 10.3. In addition, high carbon intensity, particularly in electricity production, is still evident. Although renewable energy and energy efficiency policies have been implemented in many countries in Asia, there is room for improvement toward a low-carbon path, taking into account the fact that greenhouse gas emissions in 2035 for the whole of Asia will contribute almost 50 % of global emissions (Fig. 10.4).

10.1.4 Decoupling of GHG and GDP

Entering into a low-carbon pathway means driving the country's development with low emissions of greenhouse gases. Decoupling of CO_2 emissions from GDP growth is one of the indicators showing that the path of development has to take low-carbon technology and activities into account (see Fig. 10.5). Many developing counties such as Japan, Germany, the USA, Australia, France and the UK have been through this disconnection while some prominent countries in Asia such as China, India, Malaysia and Thailand have not reached met the point of decoupling.

10.1.5 How Can Asia Leapfrog to a Low-Carbon Society?

In the situation of Asian development, green growth policies are promoted. However, looking back to the development from 1990 until 2010, as seen in Fig. 10.6,

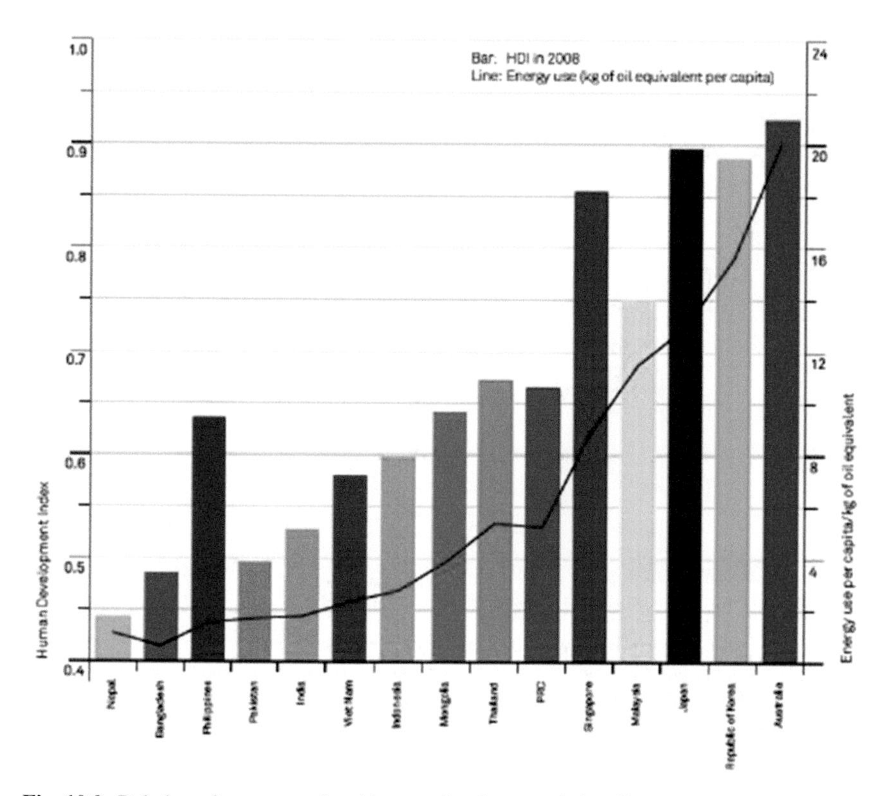

Fig. 10.3 Relation of energy used and human development index (Source: ADB 2013)

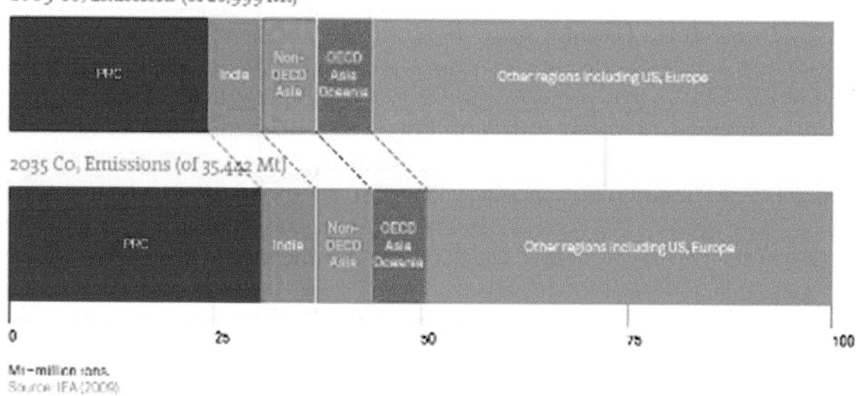

Fig. 10.4 Future emission contributions of Asia (Source: ADB 2013)

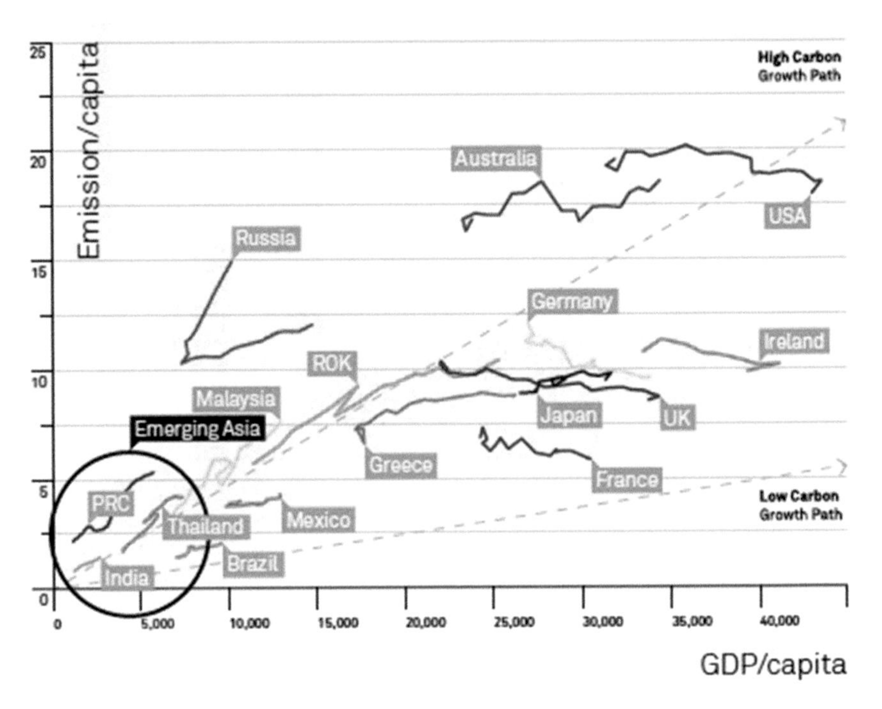

Fig. 10.5 CO_2 emissions and gross domestic product in selected countries (Source: The World Bank 2012)

real GDP growth has been increasing in parallel with CO_2 emission increases. The growth rate and emission rate have increased sharply. Conversely, population growth has increased at a slower rate when compared to other parameters. The implication of this figure is shown in Table 10.1, where the GDP of ASEAN, particularly four countries in Asia (China, India, Japan and South Korea) in the next 15 years (until 2030) will cover 38 % of the global GDP, which will be increased by almost 10 % from 2010. This potential growth in GDP is from 47 % of the world population, where its share has been constant since 2010. These constant population shares will take responsibility for the increasing GDP development of the countries. Therefore the future activities of these populations are crucial for the pursuit of low-carbon development. These activities will be integrated with technology-based and behavioral-based functions. Considering the various circumstances and the different natures of the Asian countries, comprehensive knowledge and technology transfers are essential in order to increase capacity at various levels. Understanding of climate change and its impact, as well as mitigation and abatement, are the key to initiating activity for unlocked carbon intensity development. Responsibility arising from understanding will lead to

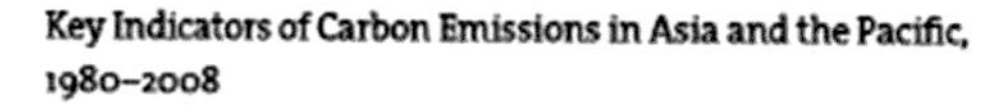

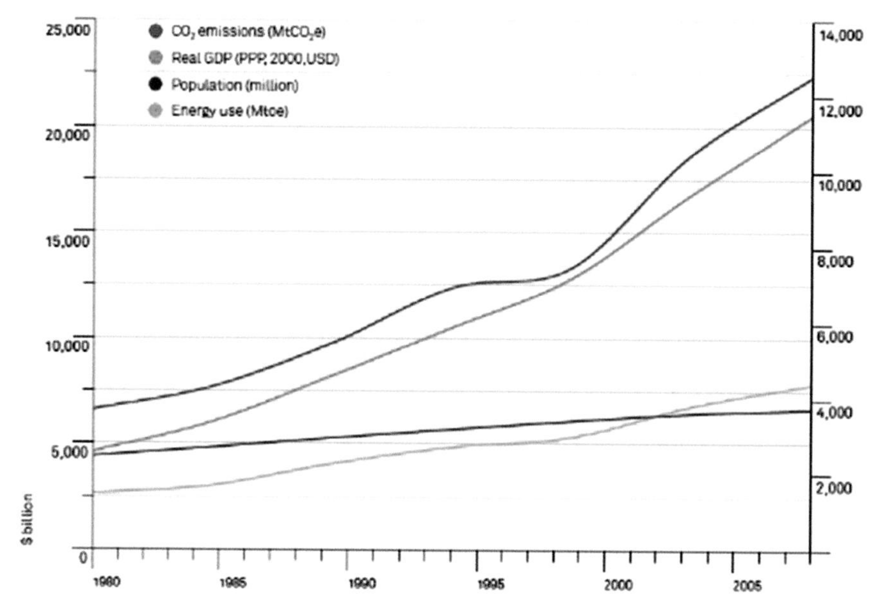

Fig. 10.6 Key indicators of carbon emissions in Asia and the Pacific (Source: ADB 2013)

Table 10.1 Changes of population and GDP between 2010 and 2030 for ASEAN + 4[a]

	Population (million)		GDP (billion USD)	
	2010	2030	2010	2030
ASEAN	592	704	1852	5587
ASEAN + 4	3206	3790	15,979	46,501
World	6641	7933	62,124	129,547
Percentage of ASEAN + 4 in the world	48 %	47 %	26 %	36 %

Source: Derived from ADB 2013, Low Carbon Green Growth in Asia
[a]+4 = China, India, Japan and South Korea

sustainable development. Capacity development to direct society to know what to do, and how to do it, is the key to success in order to leapfrog for a massive reduction in GHG.

10.2 Structure and Mechanisms of Capacity Development

In order to understand the low-carbon pathway that can help Asia unlock carbon intensity, a good structure and mechanisms of capacity building need to be designed across the region. The structure to develop capacity should have a broad space for knowledge-sharing and a full loop of knowledge transfer which starts from research and leads to policy and to various level of implementation.

10.2.1 Knowledge-Sharing Platform

Sharing of knowledge should be done on both the horizontal and vertical levels. This means that knowledge is shared among the same level of society and through different levels of the related society. There are at least four levels of society, including the community, researchers, practitioners and policy makers. To accomplish capacity development, different mechanisms are applied in each society. The 'community' itself is the ground basis for implementation of low-carbon development. Mass media and direct communication to disseminate knowledge and information through the society are good mechanisms as this is a broad and heterogeneous society. Integration of a good knowledge management system into everyday life would be an effective tool of communication. The key function of the society is understanding of the facts and impacts of climate change as well as perception of their own adaptive capacity to be resilient to global change. In Asia, this society is mostly low-income to middle-income communities and they are the majority of the Asian population. Empowering this society is an effective strategy to leapfrog toward a low-carbon society.

A society of scientists and researchers is unique in the way that knowledge and information are developed from this society. The mechanisms of research and education should be promoted in order to find real solutions to the change and to cope with the change. This society is the key to fostering capacity development in this region through knowledge management. Collaboration among Asian countries needs to be strengthened in order to gain and exchange experience as well as joint research in the framework of the Asian region as climate change is a global issue but different coping situations can be learnt from each other. Nevertheless, the key issue of this society is to connect to other levels of society in order to disseminate the knowledge to the real implementers.

Practitioners are mostly people from government and industry who are involved in implementation either by planning or actions. This community drives low-carbon activities in the real sector to be implemented. They need to be equipped with skill and to be competent to initiate activity, suggest appropriate technology and solve any problems that may occur during implementation. The mechanism to promote LCD of this society is to improve their skill and enhance technology and knowledge transfer to facilitate their activities.

The top level is policy makers where their role is planning and laying down the strategies and policy. To enhance the capacity for low-carbon development, policy makers should have special mechanisms such as high-level dialogue or high-level executive training. Nevertheless the achievement of this will lead to a high impact of change. A key issue is dissemination of the right and most feasible information in a timely manner so that good policies will be executed in a timely manner.

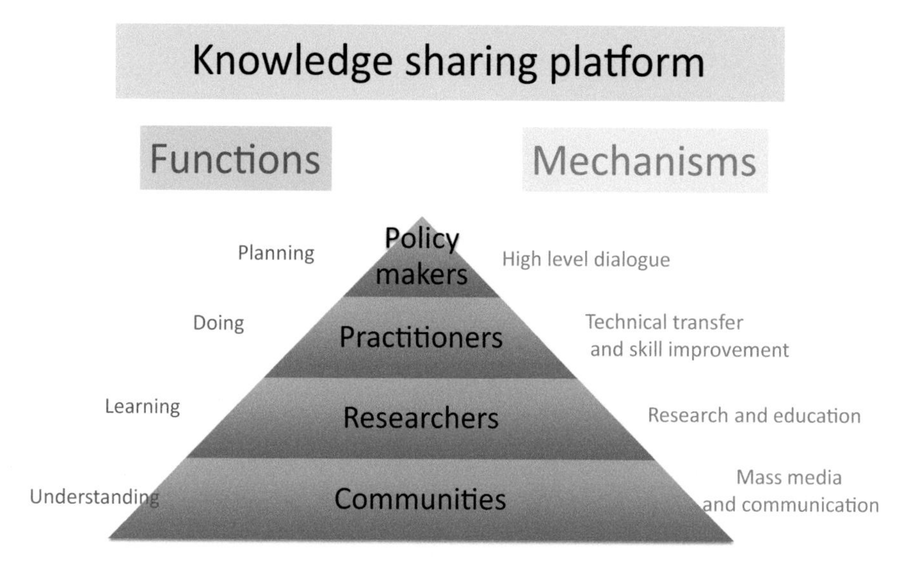

Fig. 10.7 A knowledge-sharing platform among different societies on their function and mechanisms

There is a need for Asian countries to have a knowledge-sharing platform to circulate the information at different levels of society so that they will all understand the same target and the reason for each policy and action to be pursued and implemented. Designs for connected pathways between these societies have to take into account how to stimulate the action of these sharing platforms for the greatest benefit of low-carbon development.

Figure 10.7 shows the linkage of four societies linked to the knowledge platform for both their function and mechanisms. Transforming of knowledge can be intra-society and inter-society.

10.2.2 From Research to Policy and Implementation

Not only the platform of knowledge-sharing but also the ability to make things happen both need a driving body. From research to policy, as mentioned in many scientific forums, it is not enough to unlock carbon intensity development. Figure 10.8 shows the linkage of research to policy and to implementation by the driving body in each component for clearer understanding. Usually, universities, research institutes and even non-governmental organizations that work on research are the main bodies to initiate knowledge, information and technology while the policy makers are the government itself or the ministry. Connections of the bodies

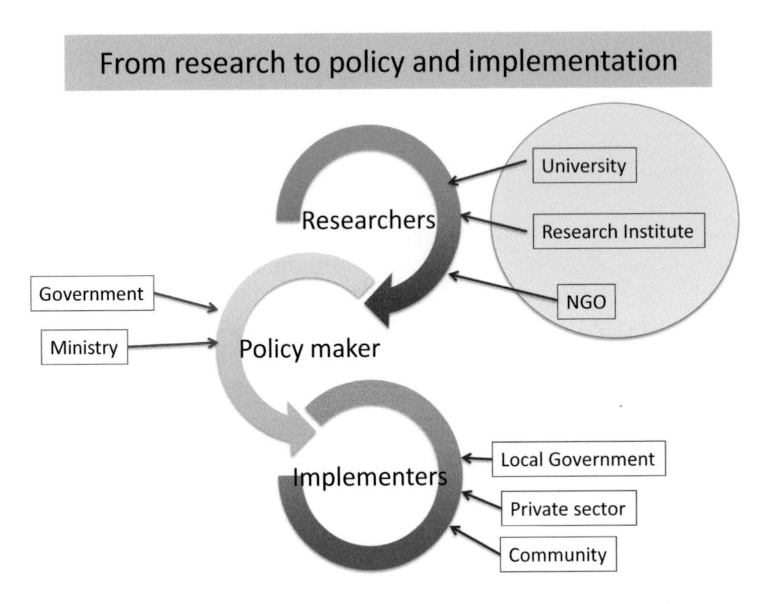

Fig. 10.8 Institutional body in the context of research, policy and implementation

of research to policy are established in some Asian countries such as Thailand where the Division of Science, Technology and Innovation Office of Planning and Policy (STI) and the Ministry of Science, Technology and Innovation initiate the MOU system with some universities and work together on how to launch the policy under the science and research support. The implementation of policy needs implementers. Local government, the private sector and communities are the different players in each role according to the policy formulation.

Institutional arrangements to cover the full loop of research to policy and to implementation can chain the different actions of the organizations mentioned above to implement low-carbon activity effectively. Any committee appointed to tackle low-carbon development should comprise the three figures of research, policy and implementation to push all action into full implementation.

10.2.3 Level of Low-Carbon Society Implementation

In order to show a good practice of capacity development that has shown potential for the leapfrog to a low-carbon society, some example of this scheme at different levels of implementation are shown below.

Community Level Ban Pred Nai Community on Environmental Protection and Energy Dependence

Ban Pred Nai is a small community located in the coastal area of the Trang Province in the eastern part of Thailand. This community, led by the Venerable Phra Subin Praneeto, who has preached the concept of truthful words which create "trust", has not only successfully preserved the local mangrove forest, but has also used improvisation from forest products and nature to generate both income and better wellbeing for the community. It is a unique knowledge transfer of the 'Sufficiency Economy Philosophy' (SEP) wisdom from this venerable monk to the community where rules and activities to protect the mangroves and energy independence have happened.

The SEP concept was introduced to the village by the Venerable Phra Subin Paneeto who endorsed the implementation of coexistence with nature to the community. In order to have an environmental management plan, the micro-credit saving fund called ' Sajja group' was operated and the revenue from this activity was raised to help community members replant mangrove trees while setting up local regulation to sustainably allow members of the community to have protection while living positively with the mangroves. The saving fund in terms of cooperation business was now operating with more than 700 million baht per year. By protecting the mangrove forest, it was shown that the Ban Pred Nai community could reclaim 2000 ha of forest back from the year 2001 onward after implementing the concept of coexisting with nature, which is one of the SEP concepts. In addition, after the mangroves had been restored, the local sea crab business was reactivated again, introducing big incomes to the community. In terms of climate change mitigation, it was found that through communal management, the mangrove forest in Ban Pred Nai absorbs 1.85 tonnes of carbon dioxide per person per year, when compared to the non-communal forest absorption rate of 0.91 ton per person per year. The reforestation activity in Ban Pred Nai creates a CO_2 sink of 2.0 tonnes of CO_2 per capita per year (Towprayoon et.al. 2011).

Not only environmental management but also implementation of SEP can help protect the environment. Being aware of the erosion of the coastal shoreline causing a loss of the mangrove area, sea water invasion and a reduced biological food chain, the Ban Pred Nai community initiated the local technology to prevent coastal erosion. A blockade made from used tires in a cubic shape has been placed along the mangrove shoreline since 1992. It was found later that this activity can protect against erosion of the shoreline, while the area of non-blockade placement failed to be maintained and collapse of mangrove trees and loss of shoreline occurred. In addition, the placement of the blockade became the routine work of the community. This indicated good understanding of self-sufficient living with self-development of the people in Ban Pred Nai. It was also a good demonstration of how the community can adapt themselves to the dimension of climate change and showed the coping capacity to deal with it.

The issue of climate change is not only related to adaptation but also to greenhouse gas mitigation. The reduction of carbon dioxide is not the major concern in SEP but to live sufficiently using fuel that can be accessed within the village is the key. Farmers in Ban Prai Nai earn their living with a mixed fruit tree orchard including rambutan, loongkang, jackfruit and durian. During the off season, farmers have to clear their excess tree branches which become wood residue. In order to avoid residue burning, several locally designed charcoal kilns have been constructed throughout the village to produce in-house use of charcoal and wood vinegar. This activity helps to produce more than 16 tonnes per year of charcoal from 53.7 tonnes per year of wood residue. This can replace around 10 tonnes of LPG, avoiding an LPG cost of approximately 200,000 baht per year. It should be noted that the community in Ban Prai Nai is only 650 people and this avoid costs of approximately 2 % of their incomes.

Subnational Level Low-Carbon City at Muang Klang Municipality

The activities of the low-carbon city in the Muang Klang Municipality in the Rayong province of Thailand are a good example of knowledge transfer from research to policy and to implementation. The Muang Klang Municipality is a member of the ICLEI but the actions of the low-carbon activities were themselves in the spotlight after a researcher from the Thailand Greenhouse Gas Organization of Thailand and the Joint Graduate School of Energy and Environment, King Mongkut's University of Technology Thonburi, set up the program with the Mayor to estimate the municipality greenhouse gas inventory from four major sectors and set up the target of reduction including energy efficiency in building, transportation, agriculture and the waste sector. The nine-steps approach to estimate GHG was initiated by JGSEE and implemented for the first time in this municipality, as seen in Fig. 10.9. The nine steps take into account the action plan where all stakeholders from the governmental office, industry and education join together and identify activities together in order to reach the target set by the study (The Joint Graduate School of Energy and Environment (2011)). Below are some action plans that have been implemented.

Mitigation Actions in the Waste Sector in the Muang Klang Municipality

- Install a municipal waste separation belt to sort organic waste and recyclables from general waste prior to landfill disposal.
- Collect fat and oil food waste from restaurants and markets to produce solid fuel used in the municipality's own slaughter house.
- Collaborate with the Ministry of Energy to install anaerobic digestion in order to produce methane gas used for heat production.

Mitigation Actions in the Agricultural Sector

- Convert unused land areas to rice fields and construct a municipal rice mill for local processing and consumption to reduce emissions from transporting rice from elsewhere.

The direct and indirect benefits from mitigation actions in the Muang Klang Municipality are shown in Table 10.2.

Regional Level Climate Change International Technical and Training Center (CITC)

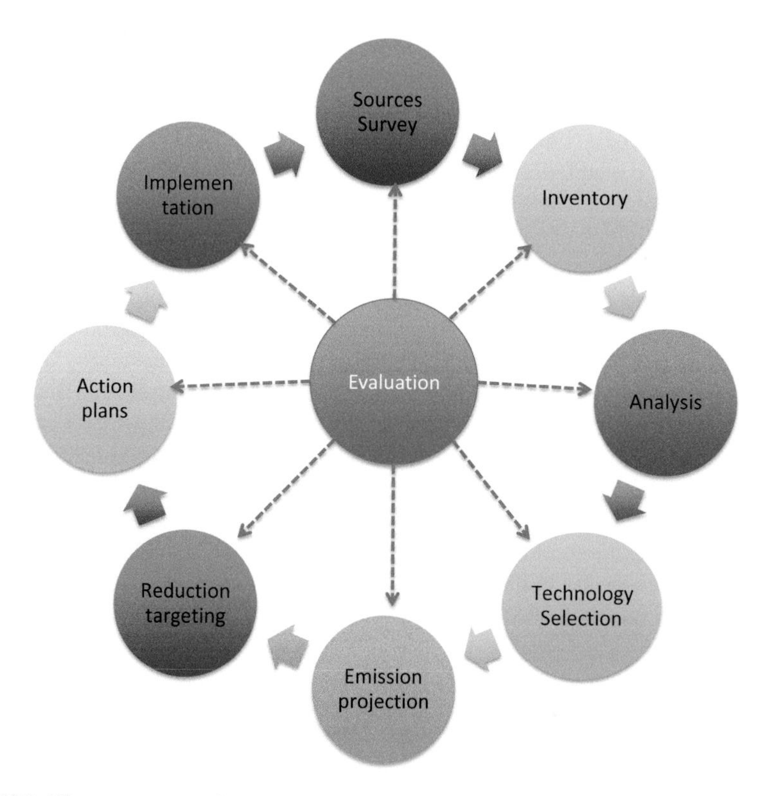

Fig. 10.9 Nine-steps approach to a low-carbon city by JGSEE

Table 10.2 GHG reduction and co-benefits

Mitigation measures	GHG emission reduction	Expected co-benefits
Installed municipal waste separation belt to sort organic waste and recyclables from general waste prior to landfill disposal	448.4 t CO_2 e avoided over 10 years from landfill methane	Lowered solid waste disposal costs for municipal authorities by 312,500 baht over lifetime of equipment (10 years)
		New revenues generated from sale of recyclables
		Extended life of municipal landfill
Constructed municipal rice mill for local processing and consumption	At least 61.6 tCO_2 e avoided from transport of rice from outside the municipality	New income generated from rice sales, benefiting smaller-scale farming households
		Reduced dependence on price in the rice market and purchases from outside the municipality
		Increased food security for local communities

There is a proposal by the Thailand Greenhouse Gas Management Organization (TGO), which is the responsible agency in Thailand for GHG mitigation activities, to establish a Climate Change International Technical and Training Center (CITC), which is aimed to be a "one-stop technical training center" and networking platform on mitigation and adaptation for ASEAN countries and other developing countries (Thailand Greenhouse Gas Management Organization 2015).

The main activities of the CITC are to provide a training service in the area of climate change mitigation and adaptation, establish a networking platform for ASEAN countries, disseminate knowledge on climate change mitigation and adaptation, and be a learning resource center on climate change mitigation and adaptation. The target groups of the CITC are governmental agencies, academic institutions, private companies related to mitigation and adaptation, and the general public. The center currently has four major courses including GHG inventory management, a mitigation mechanism, low-carbon society development and sustainable GHG management. The course are offered at different levels to include practitioners, executives and those who are interested. The CITC is supported by JICA and other international agencies including the Thai Government. It is expected that capacity building through this center will raise the standard of knowledgeable people in ASEAN.

10.3 Mechanism of Knowledge Dissemination

Capacity building at various levels can increase understanding and enhance the radius of perception but may not be enough to unlock carbon intensity over time. Dissemination of knowledge with such capacity as has been built can accelerate Asian countries' ability to fight the temperature increase of 2 °C. Knowledge dissemination can be through networks, forums or initiatives across Asia. These activities are illustrated below.

Networking Low Carbon Asia Network—LoCARNet

This network is a good example of regional collaboration in Asia and a platform to disseminate research for policy formulation.

LoCARNet is a network of researchers that facilitates the formulation and implementation of science-based policies for low-carbon development in the Asian region. It aims to facilitate science-based policies in order to realize a sustainable future based on a stabilized climate. The network endeavors to establish research capacity in the region based on South–South–North cooperation, and to reflect research findings in actual policies to achieve low-carbon growth (LCS-RNet Secretariat 2015).

With the success of the Low Carbon Research Network (LCS-RNet), the International Global Environment Study (IGES), which acted as the secretariat, has launched LoCARNet as an autonomous research network, operated through voluntary initiatives by researchers in various countries, sustaining close links with other like-minded stakeholders. LoCARNet aims to promotes research to support the development of policies for low-carbon growth by enabling dialogue between scientists and policy makers and to help the Asian region to move forward with low-carbon growth, with a number of ongoing favorable conditions to turn challenges into opportunities.

LoCARNet conducts a platform of information exchange as well as updating research progress, including pushing joint research on climate change in Asia. The uniqueness of LoCARNet is the creation of a network of prominent researchers in science, science policy and dialogue. The ownership of the knowledge is by countries and strengthening the South–South–North collaboration on climate issues.

The success of LoCARNet has been reflected by the number of country collaboration platforms, joint research projects and publications.

Research Forum Established in 2006 with the leadership of Kyoto University, Japan, and King Mongkut's University of Technology Thonburi, Thailand, the Sustainable Energy and Environment Forum (SEE Forum) is an Asia–Pacific academic and science and technological forum that brings forward dialogue on the global climate and energy security issues of common concern. The primary objective of the SEE Forum is to seek academic and science and technological cooperation that will contribute to solving the global climate and energy security issues (Sustainable Energy and Environmental Forum, 2015).

There are ten member countries—namely Brunei, Cambodia, Indonesia, India, Malaysia, Philippines, Japan, Singapore, Thailand, and Vietnam—that participate in this forum. The SEE Forum members will, either individually or collectively, bring the spirit of our common understanding and resolve to the attention of relevant policy makers and other science and academic networks, at national as well as international levels. Ultimately, the goal is to provide government agencies and policy makers with the information required to make sound decisions on global climate and energy security issues.

The main activity of the forum is to establish a research network that can work together on relevant topics with a focus on Asia. Promotion of bilateral and multilateral research, as well as exchange resources, including students, researchers and professors, are also a focus. The forum has allowed young SEE Forum researchers from Asian countries to work together on research into the energy situation in Asia. Within this initiative, knowledge of low-carbon activities has been disseminated from competent and experienced researchers to the younger generation. Members gather every year to update information and seek research collaboration. Linkage with the Asia University Network (AUN) and joint research supported by the SATREP Program are some of the products of this forum. More information can be found at www.seeforum.net.

Initiative—Global Warming Forum (Thailand Research Fund) This is an initiative by the Multilateral Environmental Agreements Knowledge and Strategic Development Think Tank Project (MEAs Think Tank) supported by a Thailand research fund (Global Warming Forum 2014). The forum frequently sets up dialogue and seminars on current topics on the low-carbon society with the right target group and consecutively updates information on low-carbon activities including negotiation and prominent scientific issues. The uniqueness of this initiative is the combination of participants comprising government, private industry, universities, NGOs and the community. This forum acts as a think tank and disseminates discussion papers, books and other output to policy makers and society.

Human Resource Development A leapfrog to a low-carbon society needs understanding of the real situation and problems in the region. This can be done collaboratively among countries in the region. An example of human development that answers the issue above is the Joint International Postgraduate Program on Energy and Environment (JIPP), which has been established by five leading universities in ASEAN, namely the Institute of Technology Bandung, Indonesia; University of Malaya, Malaysia; University of Philippines, Philippines; King Mongkut's University of Technology Thonburi, Thailand; and Hanoi University of Science and Technology, Vietnam. The program has led to a growing human resource and a network of experts who can drive energy technology and the energy market integration process in ASEAN. It will serve as a crystallization point to strengthen the cooperation in research and development among the universities in the member countries. The objectives of the program are to produce postgraduate students who can understand and find solutions to energy and environmental issues with the view of the ASEAN region and to support non-boundary competent human resources across the ASEAN region, as well as stimulating joint research and development among ASEAN universities. This program provides for students at the Masters and PhD levels in each university. The students have a chance to understand problems in other countries and collaborate on research topics under group supervisors from member universities. Mobility between universities is also encouraged (The Joint Graduate School of Energy and Environment 2014).

10.4 Conclusion and Key Messages

It is clear that the Asian countries will be an important region in the world in the future in terms of economic development and use of fossil fuels. While economic growth is projected to sharply increase by 38 % by 2030, the growth of the population is constant. Therefore, increased capacity of this region to cope with climate change is important, particularly given that the same population will drive global GDP growth for almost one third of the world. There are a number of activities currently that have shown their success and can be shared among countries. Capacity building can be done at the levels of communities, researchers, practitioners and policy makers in order to unlock from energy intensity and to leapfrog to a low-carbon society.

10.4.1 Capacity Development Is the Basis for Asia's Leapfrog

Most major development paths of Asian countries are moving toward green growth. Under the constraints of the energy crisis and climate change impact, future Asian growth, while appearing to be the most significant in comparison with other regions, needs a good knowledge-based pathway to light up and pave the road to a low-carbon society.

Capacity development is the basic need and an urgent issue to be explored in Asia. It is one of the effective tools for Asia to leapfrog to a low-carbon society with the concern of unlocked carbon intensity development.

10.4.2 A Bullet Train Model

Development needs to be done on several levels from communities to the subnational and national levels. To leapfrog from the current situation, capacities need to be built at many levels through various mechanisms of networking, research forums, initiatives, training, etc., in order to bridge, transfer and transform the results from research to policy and to implementation. Policy makers with good understanding, as the head of the bullet train, will lead society in the right direction, while scientists and researchers are the engines to back up and accelerate this movement. Finally, practitioners in communities play key roles as the fuel, enhancing the movement toward green growth through their activities. It is, therefore, essential to have these three components for a compatible basis of knowledge and comprehension through capacity development.

10.4.3 Asian Countries Need Collaboration

The Asian countries are different in nature but rich in culture and resources. Low-carbon activities are various and depend on internal factors and situations. There are many good practices and philosophies that can be shared among the countries. The experience of learning from each other facilitates accomplishments and reduce risks in implementation. Collaborative activities in capacity development help Asian countries move toward green growth in their own ways with their own uniqueness while seeing the same goal in the future together.

References

Asian Development Bank Institute (2013) Low carbon green growth in Asia: policies and practices. A Joint Study of the Asian Development Bank and the Asian Development Bank Institute. ISBN 978-4-89974-037-3

Global Warming Forum (2014) MEAs Watch. http://www.measwatch.org. Accessed 9 Dec 2014 (in Thai)

Joint Graduate School of Energy and Environment, KMUTT (2011) A study submitted to Thailand Greenhouse Gas Management Office. Thai pilot study of low carbon city in Muang Klang Municipality

LCS-RNet Secretariat (2015) About LoCARNet. http://lcs-rnet.org/about_locarnet/. Accessed 15 Jan 2015

Sustainable Energy and Environmental Forum (2015) Introduction. www.seeforum.net. Accessed 15 Jan 2015

Thailand Greenhouse Gas Management Organization (2015) CICT in brief. http://citc.in.th/index.php/en/. Accessed 20 Jan 2015

The Joint Graduate School of Energy and Environment (2014) Joint international postgraduate program on energy and environment http://www.jgsee.kmutt.ac.th/jgsee1/researchProject/jipp/132.pdf. Accessed 1 Dec 2014

The World Bank (2012) World Development Indicator (2012). ADB-ADBI Study Team. http://data.worldbank.org/data-catalog/world-development-indicator

Towprayoon S, Kadkarnkai Y, Srethasirote B, Sathirathai S (2011) An approach to sufficiency society: a case study of Ban Pred NaiCommunity. In: Fostering economic growth through low carbon initiatives in Thailand, Chula Global Network, Chulalongkorn University

Chapter 11
Capacity Development on GHG Inventories in Asia

WGIA Workshop on Greenhouse Gas Inventory in Asia

Hiroshi Ito

Abstract The Greenhouse Gas Inventory Office (GIO) of Japan has organised the "Workshop on Greenhouse Gas Inventories in Asia (WGIA)" since 2003. The workshop is tasked to improve GHG inventory dataset credibility in Asia and help bind countries within the Asian region. Participating countries are Cambodia, China, India, Indonesia, Japan, the Republic of Korea, Lao PDR, Malaysia, Mongolia, Myanmar, the Philippines, Singapore, Thailand and Vietnam (14 countries). Since the 6th WGIA (WGIA6) in 2008, WGIA has been convened as part of the "Kobe Initiative" of the G8 Environment Ministers' Meeting. WGIA participants are government officials, inventory compilers, researchers and staff in international organisations. The workshops have been held in other Asian countries to help attract more attendees. Participants from many countries can conduct face-to-face discussions at WGIA. Many achievements were realised through the workshops, in particular:

- Establishment of the WGIA network platform to exchange information on climate change and mitigation of GHG emissions as well as GHG inventory
- Sharing of information and experiences that can be beneficial for other countries
- Identifying common problems and possible solutions
- Updating of the status of national inventory development

This collaborative approach may be applicable for other regions.

H. Ito (✉)
National Institute for Environmental Studies, Ibaraki, Japan
e-mail: ito.hiroshi@nies.go.jp

S. Nishioka (ed.), *Enabling Asia to Stabilise the Climate*,
DOI 10.1007/978-981-287-826-7_11

Keywords WGIA • GHG inventory • Biennial update reports • Nationally appropriate mitigation actions • MRV • GIO of Japan • Capacity development • Mutual learning

Key Message to Policymakers

- GIO has conducted a Workshop on Greenhouse Gas Inventories in Asia (WGIA) annually for 12 years.
- Workshop continuity helps develop networks.
- WGIA operates to exchange information among inventory experts.
- Face-to-face workshops are necessary for developing relationships of mutual trust.

11.1 Introduction to WGIA

11.1.1 GHG Inventory in International Negotiations

The 5th Assessment Report published by the Intergovernmental Panel on Climate Change (IPCC) in 2013 stated that "the atmospheric concentrations of the greenhouse gases carbon dioxide (CO_2), methane (CH_4), and nitrous oxide (N_2O) have all increased since 1750 due to human activity".

In accordance with Articles 4 and 12 of the United Nations Framework Convention on Climate Change (UNFCCC), all Parties to the Convention are required to submit greenhouse gas inventories to the Conference of the Parties (COP) under the convention as part of their national communications (NCs) at a frequency determined by the COP.

GHG inventories are important for ensuring the transparency and accuracy of each country's mitigation actions by quantifying anthropogenic GHG emissions. In this respect, national GHG inventories, which provide information on the GHG emissions and their trends over time, play a critical role as a basis for decisionmakers to design and implement strategies for mitigation actions and GHG emission reductions within their country.

Inventories form the basis of national policy development because they can be used to:

- Identify the major sectors where abatement will have a real impact.
- Predict and compare impacts of mitigation measures.
- Choose cost-effective options.

Inventories are essential to monitoring of impacts of mitigation policies and measures because:

- Policymakers need to know if policies are working.
- They need to reflect impacts of mitigation actions and thus require careful choice as regards method.

11.1.2 Responsibility of Developing Countries

Only developed countries, which are the main emitters of greenhouse gas, have climate change responsibilities and are mandated to create and submit GHG inventories periodically. However, recently, developing countries have achieved rapid growth, and the emissions of greenhouse gas caused thereby have significantly increased, which means developing countries will also have to start submitting GHG inventories. It is thus necessary for each country to urgently asses its national circumstances.

Since the Bali Action Plan, which states that Non-Annex I Parties should also take nationally appropriate mitigation actions in a measurable, reportable and verifiable manner, was agreed on at COP13 in 2007, the importance of greenhouse gas inventories has been recognised as a tool for supporting the developing mitigation measures and to verify their efficacy. From the Cancun Agreements, Non-Annex I Parties shall make the biennial update report every 2 years (see Table 11.1).

Table 11.1 Biennial update report, Decision 1/CP.16—Cancun Agreements (Reference: UNFCCC (2011))

	Frequency	Content
National communications	4 years	National circumstances
		GHG inventory
		Adaptation and mitigation action
		Relevant information
		Necessary support
Biennial update reports	2 years	GHG inventory
		Information on mitigation action
		Needs and support received

11.1.3 The Role of Greenhouse Gas Inventory Office of Japan (GIO)

The Greenhouse Gas Inventory Office of Japan (GIO) was established in July 2002 in the Center for Global Environmental Research (CGER) at the National Institute for Environmental Studies (NIES). Its mission is to compile the annual national greenhouse gas (GHG) inventory of Japan; to implement various GHG inventory-related tasks and activities, such as providing support and assistance for the technical review of the national GHG inventory of Japan for the UNFCCC and the Kyoto Protocol; and to contribute to capacity building of Asian countries in developing and improving their GHG inventories (see Table 11.2).

The "National GHGs Inventory Report of JAPAN (NIR)" and "GHGs Emissions Data of Japan", both of which are published annually, as well as information on and reports from the "Workshop on Greenhouse Gas Inventories in Asia (WGIA)" are available and posted on the GIO website.

Additionally, some members join the process of the technical review of other parties for the UNFCCC and the Kyoto Protocol in some countries such as Germany as a member of the expert review team (ERT).

11.1.4 One Part of the National System

The Ministry of the Environment of Japan (MoEJ), with the cooperation of relevant ministries, agencies and organisations, prepares Japan's national inventory and compiles supplementary information required under Article 7.1, which is annually submitted to the Conference of the Parties through the UNFCCC Secretariat in accordance with the UNFCCC and the Kyoto Protocol.

The MoEJ takes overall responsibility for the national inventory and therefore does its utmost to improve the quality thereof. The MoEJ organised the "Committee for the Greenhouse Gas Emission Estimation Methods" in order to integrate the latest scientific knowledge into the inventory and to ensure it reflects recent

Table 11.2 List of tasks of Greenhouse Gas Inventory Office of Japan (GIO)

Task of Greenhouse Gas Inventory Office of Japan
- Preparing annual national GHG inventory
- Providing support for Japan's national GHG inventory
- Support and assistance of political actions relating to GHG inventory
- Convening the Workshop on GHG Inventories in Asia (WGIA)
- International cooperation for improvement of GHG inventory
- Participation in GHG inventory review as reviewer

international provisions. The estimation of GHG emissions and removals, the key category analysis and the uncertainty assessment are then carried out by taking the decisions of the committee into consideration. Substantial activities, such as the estimation of emissions and removals and the preparation of Common Reporting Formats (CRF) and National Inventory Report (NIR), are performed by the Greenhouse Gas Inventory Office of Japan (GIO), which belongs to the Center for Global Environmental Research of the National Institute for Environmental Studies. The relevant ministries, agencies and organisations provide the GIO with the appropriate data (e.g. activity data, emission factors, GHG emissions and removals) through compiling various statistics and also provide relevant information on supplementary information required under Article 7.1. They then check and verify the inventories (i.e. CRF, NIR), including the spreadsheets that are actually utilised for the estimation, as a part of the quality control (QC) activities.

The checked and verified inventories determined as Japan's official values are then published by the MoEJ and submitted to the UNFCCC Secretariat by the Ministry of Foreign Affairs (Reference: Ministry of the Environment, Japan and Greenhouse Gas Inventory Office of Japan (GIO), CGER, NIES (2014)).

Figure 11.1 shows the overall institutional arrangement for Japan's inventory preparation.

11.1.5 The Objective of WGIA

Thus far, on the basis of Articles 4 and 12 of the UNFCCC, Annex I countries have compiled a GHG inventory annually, but Non-Annex I Parties have only done so once or twice and with the national communications (NCs).

However, at COP16 in 2010 and COP17 in 2011, it was agreed that, in addition to the NCs, all Parties to the Convention, including Non-Annex I Parties, shall submit information on GHG inventories as a biennial update report (BUR). It was also agreed at COP17 that developing country Parties should submit their first BUR by December 2014 and subsequent reports every 2 years. For this reason, more accurate inventories, which support the development of mitigation measures and the verification of the effectiveness of these measures, need to be reported at a higher frequency than ever before. The importance of periodical GHG inventories

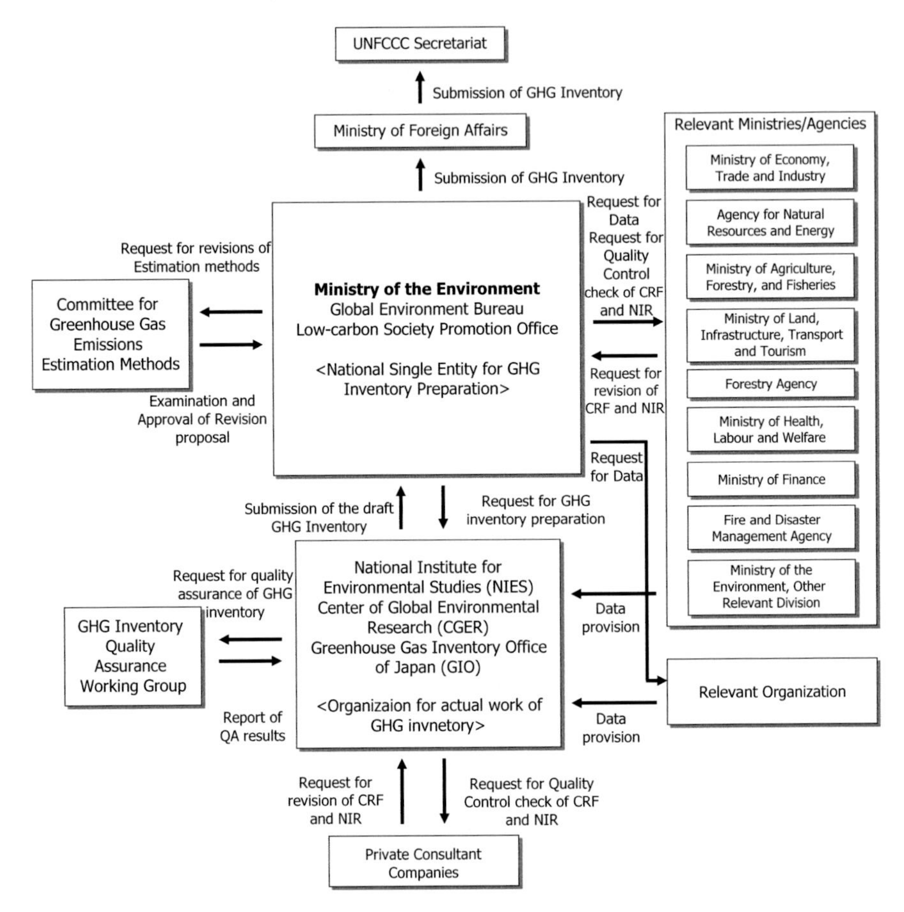

Fig. 11.1 Japan's institutional arrangement for national inventory preparation (Reference: Ministry of the Environment, Japan and Greenhouse Gas Inventory Office of Japan (GIO), CGER, NIES (2014))

is increasing on an international basis, which means Non-Annex I Parties unfamiliar with compiling the GHG inventory periodically will require capacity building.

With this aim in mind, GIO convened the Workshop on Greenhouse Gas (GHG) Inventories in Asia (WGIA) in 2003. In order to improve the quality of GHG inventories, it is important for the related countries to exchange information on them, as this will aid inventory compilers and administrators managing the compilation. Since 2003, government officials, inventory compilers and researchers directly involved with inventory preparation in the participating countries have met to exchange information at the workshop (Reference: Greenhouse Gas Inventory Office of Japan (GIO), CGER, NIES (2012)).

The objectives of the workshop are:

- To enhance sector-specific capacity for inventory compilation (mutual learning)
- To facilitate periodical national GHG inventory preparation for national communications (NCs) and biennial update reports (BURs)

- To discuss the possibility of inventories as a supporting tool for mitigation measures/NAMAs
- To explore issues on measurability, reportability and verifiability (MRV) at various levels
- To provide an opportunity for countries in the Asian region to cooperate and share information and experiences related to their own national GHG inventories
- To support countries in Asia in improving the quality of inventories via regional information exchange (Reference: Greenhouse Gas Inventory Office of Japan (GIO), CGER, NIES (2015))

Participating countries are Cambodia, China, India, Indonesia, Japan, the Republic of Korea, Lao PDR, Malaysia, Mongolia, Myanmar, the Philippines, Singapore, Thailand and Vietnam (14 countries).

WGIAs consist of the following sessions:

- Plenary sessions
- Sectoral working group sessions
- Mutual learning sessions
- Hands-on training sessions

Basically, each WGIA consists of three of the above four sessions.

11.1.6 History of WGIA

Japan is the only Annex I Party in Asia with experience in completing the periodical GHG inventory in Asia. Since Asian cultures and climates vary greatly from those of Europe and the USA, so do the methods of estimating emissions and removals and institutional arrangements. As Japan has constructed an appropriate methodology and institutional arrangement based on Asian culture and climate, it can share its GHG inventory information through WGIA with other countries in Asia, due to their cultural and climatic similarities.

The WGIA and capacity building for measurability, reportability and verifiability were both initiated in 2003 with the aim of building capacity within Asia to develop a GHG inventory. Since its sixth meeting in 2008 (WGIA6), WGIA has been convened as a part of the "Kobe Initiative" of the G8 Environment Ministers' Meeting (Reference: Greenhouse Gas Inventory Office of Japan (GIO), CGER, NIES (2009a)).

WGIA has grown since its first meeting in 2003, from 27 participants to over 100 at WGIA7 in 2009 (Reference: Greenhouse Gas Inventory Office of Japan (GIO), CGER, NIES (2009b)) then to 130 at WGIA10 in 2012. The 2014 WGIA (WGIA12) was attended by 123 persons from 14 WGIA member countries and international organisations and is now one of the biggest climate change events in Asia. It has also received requests from countries such as Pakistan and East Timor, which are not currently members of WGIA, to attend future WGIA meetings. There is the possibility of expanding the scale of the workshop.

Table 11.3 List of host countries

Month, year	Host country	Theme
November 2003	WGIA1: Thailand	Identified problems and needs of support
February 2005	WGIA2: China	Shared information and experiences gained through inventory development
February 2006	WGIA3: Philippines	Discussed technical matters on each sector inventory
February 2007	WGIA4: Indonesia	Organised working groups and discussed sector-specific issues
September 2007	WGIA5: Malaysia	Identified needs for further inventory improvement
July 2008	WGIA6: Japan	Reaffirmed the importance of inventory development
July 2009	WGIA7: Republic of Korea	Shared information and experiences/discussed sector-specific and crosscutting issues
July 2010	WGIA8: Lao PDR	Shared information and experiences/discussed sector-specific and crosscutting issues
July 2011	WGIA9: Cambodia	Initiated "mutual learning"
July 2012	WGIA10: Vietnam	Shared information and experiences/conducted mutual learning
July 2013	WGIA11: Japan	Shared information and experiences/conducted mutual learning
August 2014	WGIA12: Thailand	Shared information and experiences/conducted mutual learning

The participating countries that acted as host countries for WGIA from 2003 to 2014 are shown in Table 11.3.

11.1.7 Contents of WGIA

1. Topics Discussed in Plenary Sessions

   The topics of discussion covered various categories on WGIA as shown below. WGIAs consist of the following sessions:

- Plenary sessions
- Sectoral working group sessions
- Mutual learning sessions
- Hands-on training sessions

Basically, as mentioned above, each WGIA consists of three of the above sessions. All participants join the plenary sessions and then choose sectoral working-group sessions as well as hands-on training sessions. Mutual learning sessions are closed sessions and are limited in participant number.

Plenary sessions deal with overall and cross-cutting issues on national GHG inventory preparation, such as data provision, institutional arrangements and introduction to countermeasures for climate change of Japan and host countries, as well as mitigation action such as NAMAs. Through discussions in the plenary sessions, participants in WGIAs share information from various data sources, which is useful for improving their inventory preparation systems.

The topics in plenary sessions were:

- Progress report on Non-Annex I Parties' national communications (NCs) shared by the UNFCCC Secretariat
- Progress of NCs and BURs in each participating country
- National systems for periodical national GHG inventory preparation
- Relationships between inventory and mitigation measures/NAMAs
- Enhancement of network for supporting measurability, reportability and verifiability (MRV)
- Quality assurance/quality control (QA/QC)
- Uncertainty assessment
- Time-series consistent estimates, etc.

In plenary sessions, UNFCCC provides information on the international framework and COP decisions. Participants welcome this presentation as cross-cutting issue such as QA/QC; UA and time-series consistency important for quality improvement of the GHG inventory are discussed.

2. Sectoral Working Group Sessions

Regarding the GHG inventory, there are many sectors and categories, such as energy, industrial process, agriculture, LULUCF (land use, land-use change and forestry) and waste. WGIAs provide the sectoral working group sessions in order to discuss particular sector-specific issues and find solutions to them. There are various issues for inventory preparation in each sector, and the sectoral working group sessions deal with sector- or category-specific issues.

Table 11.4 shows main topics containing sectoral working group sessions. The WGIA participants are government officials and inventory compilers or researchers directly involved with inventory preparation. Inventory compilers and researchers attended the breakout sessions for each expert sector or category; and government officials attended the breakout group of cross-cutting issues such as regional and/or city-level GHG inventories. Discussion of such sector-specific issues among sectoral experts is recommended in order to cover the issues thoroughly.

Table 11.4 Topics of sectoral working group sessions

	Crosscutting	Energy	Agriculture	LULUCF	Waste
WGIA6	Awareness raising of GHG inventories		Strategies to improve reliability of data	Use of remote-sensing data	Strategies to improve reliability of data
WGIA7		Statistics for energy sector	Emission factors utilized for NCs	Activity data from remote-sensing and GIS	Improvement of data collection scheme
WGIA8	Institutional arrangements for inventory preparation		Estimation methods and development of parameters	Follow up of WGIA7 (remote sensing and GIS data)	Information exchange on the current status of sectoral inventory preparation
WGIA9	Non-CO_2 gas estimation QA/QC systems	Estimation of CO_2 emissions from transport sector			Development of waste statistics
WGIA12	GHG inventory at various levels		Relationship between national GHG inventories and mitigation measures, specifically NAMAs		

3. Mutual Learning Sessions

The mutual learning (ML) session is an activity to improve the inventories of individual countries through the following processes: (1) exchanging inventories between two countries, (2) learning from a partner's inventory and (3) exchanging comments on each other's inventories. The primary purpose of ML is to improve GHG inventories by providing details of methods and data for GHG emission/removal estimation between two countries and exchanging comments on the methods and data. Studying a partner country's inventory and discussing it with its compilers provide useful information for inventory preparation and compilation. ML is also expected to foster and strengthen cooperation among GHG inventory experts in Asia. Since the aim of ML is not criticism or auditing, participants can freely communicate on a one-to-one basis as equals, rather than in one-way communication as is found with the examiner–examinee relationship (Reference: Greenhouse Gas Inventory Office of Japan (GIO), CGER, NIES (2015)).

ML was introduced to other participating countries in WGIA8 in 2010 and participants requesting mutual learning sessions between WGIA countries in the sessions of WGIA. Therefore, ML has been conducted since WGIA9 (Reference: Greenhouse Gas Inventory Office of Japan (GIO), CGER, NIES (2010)).

The ML sessions are closed sessions in order to ensure confidentiality of discussions in the sessions; only participants, chairpersons, facilitators and rapporteurs for each ML session and the WGIA Secretariat are allowed to enter conference rooms for the sessions in principle (Reference: Greenhouse Gas Inventory Office of Japan (GIO), CGER, NIES (2015)).

Through the discussions, participants studied their partner country's methodologies for GHG emission estimations, which usually differ from their own, to receive hints on improving their own inventory. They also shared any technical issues (e.g. data collection, adoption of emission factors, national system) in order to better overcome them.

Several participants in past MLs stated that they had improved their inventory through the ML experience and in particular were able to refine their inventories before official submission to the UNFCCC such as NCs and BURs. The participants in WGIA11 acknowledged the efficacy of ML in improving their inventories and agreed that implementation of MLs should continue in future WGIAs.

In the case of WGIA12, the WGIA Secretariat notified the participants of WGIA of the ML and received applications from 29 teams from eight parties on December 2013. Considering the requirements of the applicants and an appropriate balance among sectors and feasibility of implementation, the WGIA Secretariat (GIO) organised them into pairs (Indonesia and Myanmar on energy sector, China and Mongolia on agriculture sector and Vietnam on LULUCF sector) on April 2014 (Reference: Proceedings of WGIA12 2014).

Thus, the ML sessions were conducted for the energy sector, agriculture sector and LULUCF sector, as shown in Table 11.5. Participating countries studied worksheets for emission estimates and methodology reports to estimate the emissions of partners and exchanged comments and answer sheets before the WGIA discussion. Many findings and hints to improve the GHG inventories were exchanged across the table in the session of WGIA12 in Bangkok (Reference: Greenhouse Gas Inventory Office of Japan (GIO), CGER, NIES (2015)).

Prior to WGIA12, only ten countries had attended the ML sessions. As mentioned above, ML is useful for improving one's own inventory and is considered a form of external quality assurance activity by some participants. It is hoped that more participants will join the ML sessions in future WGIA.

Table 11.5 List of countries participating in mutual learning

	Energy	Industrial processes	Agriculture	LULUCF	Waste
WGIA9	Indonesia–Mongolia			Lao PDR–Japan	Cambodia, Indonesia, RoK
WGIA10	Cambodia–Thailand	Indonesia–Japan	Indonesia–Vietnam		China–RoK
WGIA11	Lao PDR–Thailand		China–Myanmar		Malaysia–Vietnam
WGIA12	Indonesia–Myanmar		China–Mongolia	Vietnam	

Reference: Greenhouse Gas Inventory Office of Japan (GIO), CGER, NIES (2015)

Table 11.6 List of hands-on training sessions

	Topic
WGIA6	How to implement a key category analysis
WGIA7	How to fill data gaps
WGIA8	How to implement mutual learning for national GHG inventories
WGIA10	How to use the new IPCC Inventory Software (energy, industrial processes, waste)

4. Hands-On Training Sessions

Most WGIA participating countries have insufficient experience in GHG inventory preparation, especially in terms of technical issues such as key category analysis and IPCC Inventory Software. Technical issues on how to implement inventory preparation obviously need addressing with training, which is why WGIA is useful as it provides hands-on training sessions. In the sessions participants can attempt to actually implement some of the technical processes of inventory preparation.

Table 11.6 shows the topics of hands-on training sessions.

11.1.8 Latest Workshop on GHG Inventories in Asia (WGIA12), 2014

The Ministry of the Environment of Japan (MoEJ) and the National Institute for Environmental Studies (NIES) convened the WGIA as a capacity building workshop for measurability, reportability and verifiability (MRV) as a part of Japan's assistance for developing countries. Ever since 2003, the workshops have aimed at supporting Non-Annex I (NAI) Parties in Asia to develop and improve their GHG inventories.

In August 2014, the 12th workshop was held and attended by over 120 experts from 14 WGIA member countries (Cambodia, China, India, Indonesia, Japan, the Republic of Korea, Lao PDR, Malaysia, Mongolia, Myanmar, the Philippines, Thailand, Singapore and Vietnam), as well as representatives from the Secretariat of the UNFCCC, Technical Support Unit from the IPCC Task Force on National Greenhouse Gas Inventories (IPCC TFI TSU), the Regional Capacity Building Project for Sustainable National Greenhouse Gas Inventory Management Systems in Southeast Asia (SEA GHG Project), the United Nations Environment Programme (UNEP), the Food and Agriculture Organization of the United Nations (FAO), the Global Forest Observations Initiative (GFOI), the Asia-Pacific Network for Global Change Research (APN), the US Agency for International Development (USAID), the US Environmental Protection Agency (USEPA) and relevant Japanese institutes in Bangkok, Thailand.

The GIO (Secretariat of the WGIA12) both organises the programmes of WGIA, according to the needs and requests of the participants, and conducts the WGIA.

In WGIA12, the biennial update report (BUR) to be submitted by Non-Annex I countries by year end and the international consultation and analysis (ICA, part of BUR) were key topics on the discussion agenda. Also discussed were the importance of accurate GHG inventories and QA/QC activities; the importance of MRV at various levels, such as region and city levels, for verification of implementation and planning for NAMA; the necessity of consolidating stable systems of GHG inventory for applying high cost–benefit technology in the AFOLU sector; and the need to maintain ongoing correspondence with inventory compilers and researchers providing new technology.

Through WGIA12, the capacity development of participating countries for MRV and the network for BUR were enhanced, with the aim of creating BUR, conducting ICA and implementing the intended nationally determined contributions (INDCs).

WGIA 13 will be held in Indonesia, where BURs submitted by Non-Annex I Parties this year will be presented by the participants. Further, mutual learning and discussions concerning ICA will be conducted. (Reference: Greenhouse Gas Inventory Office of Japan (GIO), CGER, NIES (2015)).

11.2 Achievements of WGIA

11.2.1 Enhanced Relationships

The Workshop on Greenhouse Gas Inventories in Asia (WGIA) has been run since 2003 to provide an opportunity for countries in the Asian region to cooperate and share information and experiences related to the development of the national GHG inventory. In 2014, the WGIA12 was held in Bangkok, Thailand.

As described above, Japan, the only Annex I Party in Asia, has been sharing its experiences concerning compiling the periodical GHG inventory with WGIA participants, and the participants have been sharing information related to methodology, such as country-specific emission factors for Asian countries. Since the IPCC default emission factor was not appropriate for the climate of SE Asia, particularly for agriculture, LULUCF and waste, sharing specific regional emission factors is beneficial, and in this respect, Japanese researchers provided much data to assist in the development of regional- and country-specific emission factors. Governmental officials also shared information concerning institutional arrangements based on Asian culture, and this sharing of information ensures that the methodology and institutional arrangements of Asian countries are appropriate. Building a tighter network of Japanese researchers and Asian government officers and researchers is important for the GHG inventory, as well as for countermeasures against climate change.

As mentioned above, the first WGIA in 2003 had 27 participants, which rose to 130 in 2012. The latest WGIA (WGIA12) in 2014 had a participation of 123, from 14 WGIA member countries and international organisations in 2014. WGIA has become one of the biggest events on climate change in Asia. Requests have even been received from non-member countries, such as Pakistan and East Timor, to join future workshops. As regards the size of the event, in theory it could be scaled but could suffer due to insufficient budget or capacity of GIO, the WGIA Secretariat.

In the beginning, the main participants were researchers, and topics concerned the national system and technical issues of each expert. Recently though, the proportion of government officials attending has been increasing. At the latest WGIA, not only GHG inventory technical issues but also mitigation issues and regional- or city-level inventories were discussed. Many government officials and policymakers also evaluated measurements concerning climate change.

Further, advanced research and development on emission factors and climate change issues in Japan have also been introduced, the research of which has been helpful in creating the GHG inventory for Asian countries. The introduction of climate change research in SE Asia has enabled collaboration between Japanese researchers and local researchers in other Asian countries. Japanese researchers became aware of the needs of WGIA countries through discussions at WGIA. Furthermore, WGIA also enables government officials to access the latest information on climate change research, which illustrates the importance of the government–research relationship.

Relationships between researchers and government officials are bolstered at the GIO-held WGIA every year. Further, activities unrelated to WGIA have also been held, such as the initiation of mutual learning between Japan and Korea. Mutual learning is an opportunity to understand all the different GHG inventories and how they contribute to improving GHG inventories. Korea also mentioned that mutual learning is implemented as a form of external quality assurance in the WGIA sessions. As already described in Sect. 11.1.7, mutual learning has also been conducted between other countries in WGIA sessions every year. Lao PDR, which attended the mutual learning sessions in WGIA9 and WGIA11, also introduced a mutual learning programme that emphasises peer reviews of the LULUCF with GIO. Lao PDR commented that this enhanced both accuracy and completion of the inventory of the LULUCF sector of the Lao PDR.

WGIA is financed from a budget of the Ministry of the Environment, Japan. GIO, part of the National Institute for Environmental Studies, convened the WGIA and invites researchers to discuss the technical issues of GHG inventories free of international opinion or negotiations, an environment deliberately fostered so that researches can speak freely without being hindered by governmental or international bias. This forum for free discussion was built on a relationship of mutual trust, and as it moves from country to country every year and is not solely based in Japan, this enables host countries to participate more easily. As a result WGIA can be attended by many participants, enabling face-to-face contact crucial to carrying issues forward.

11.2.2 Sharing Information Such as Sector-Specific Issues and General Issues of GHG Inventory

1. International Negotiation

TSU and UNFCCC have attended WGIA since its inception, where they continue to disseminate information on the status of international negotiations and UNFCCC mandates based on the latest information on COP. WGIA also gave government officials a chance to catch up on progress in international negotiations, and the Q&A session provides a chance to better understand institutional arrangements and policy measures.

In the actual workshop, Japan and the host countries introduce countermeasures individually taken for climate change and participating countries share their NCs. Through such presentations, progress in countermeasures for climate change of WGIA countries—which share similar climatic, international position and economic circumstances—can be shared, thus clarifying the status of each country. WGIA is thus an effective means by which to evaluate the results of policy.

2. Sharing of Information and Experience

In WGIA, current internationally relevant information and estimation methodology are discussed, which benefits other countries. Further, common problems and possible solutions are identified.

11.2.2.1 Estimation of Time-Series GHG Emissions/Removals

Mongolia estimated its annual time-series GHG emissions and removals from 1990 to 2006, as can be found at http://www-gio.nies.go.jp/wgia/wg7/pdf/4.2.5.%20Dorjpurev%20Jargal.pdf

Thailand estimated its quadrennial time-series GHG emissions and removals from 1990 to 2003, as well as annual time-series GHG emissions excluding LULUCF from 2000 to 2005, which can be found at http://www-gio.nies.go.jp/wgia/wg7/pdf/4.2.6.%20Sirintronthep%20Towprayoon.pdf http://www-gio.nies.go.jp/wgia/wg6/pdf/3-3%20Sirintornthep%20Towprayoon.pdf

Indonesia estimated its annual time-series GHG emissions and removals from 2000 to 2005, as follows: http://www-gio.nies.go.jp/wgia/wg7/pdf/4.2.7.%20Rizaldi%20Boer.pdf

11.2.2.2 Development of Country-Specific Emission Factors

China developed country-specific emission factors for CH_4 emissions from paddy fields and N_2O emissions from cropland, which can be found at:

http://www-gio.nies.go.jp/wgia/wg10/pdf/2-2_5_AFOLU_China.pdf

India developed country-specific emission factors for CH_4 emissions from enteric fermentation by ruminant animals and N_2O emissions from agricultural soils:

http://www-gio.nies.go.jp/wgia/wg8/pdf/3-wg2-3_sultan_singh.pdf
http://www-gio.nies.go.jp/wgia/wg8/pdf/3-wg2-5_chhemendra_sharma.pdf

Indonesia developed country-specific emission factors for CH_4 emissions from rice cultivation:

http://www-gio.nies.go.jp/wgia/wg7/pdf/4.2.7.%20Rizaldi%20Boer.pdf

11.2.2.3 Establishment of National Systems for National GHG Inventory Preparation

Mongolia appointed the National Agency for Meteorology, Hydrology and Environment Monitoring as its designated professional authority for national GHG inventory preparation and structured its national system, in which the agency plays the central function, information on which can be found at:

http://www-gio.nies.go.jp/wgia/wg8/pdf/3-wg1-2_batimaa_punsalmaa.pdf

Korea established the GHG Inventory & Research Center of Korea (GIR) and improved existing national system by entrusting the GIR to act as central coordinator, as explained at:

http://www-gio.nies.go.jp/wgia/wg7/pdf/4.1.5.%20Jang-won%20Lee.pdf
http://www-gio.nies.go.jp/wgia/wg9/pdf/3-wg4-4_mihyeon_lee.pdf

Indonesia enacted Presidential Regulation 71/2011 as the foundation for Indonesian GHG inventory preparation and established a national GHG inventory system; see the following for more details:

http://www-gio.nies.go.jp/wgia/wg10/pdf/3_1.pdf

11.2.2.4 Development of Quality Assurance/Quality Control (QA/QC) System

Mongolia established a QA/QC plan for energy and industrial process sector, as explained at:

http://www-gio.nies.go.jp/wgia/wg9/pdf/3-wg4-3_dorjpurev_jargal.pdf

Korea developed a QA/QC system for the waste sector and applied bilateral peer reviews its GHG inventory with Japan as one of its QA activities, as explained at:

http://www-gio.nies.go.jp/wgia/wg9/pdf/3-wg4-4_mihyeon_lee.pdf

11.2.3 Related Activities and International Cooperation

1. SEA Project

The Regional Capacity Building for Sustainable National Greenhouse Gas Inventory

Management System in Southeast Asia (SEA GHG Project) was held back to back with WGIA every year. The project is ran with the UNFCCC as the lead agency and in collaboration with US Environmental Protection Agency (US EPA), US Agency for International Development (USAID), Colorado State University (CSU), Workshop on GHG Inventories in Asia (WGIA (GIO/NIES)) and USAID Low Emissions Asian Development (LEAD) programme. The participants of the SEA GHG Project and WGIA have the same aim, and holding similar activities at the same time has a synergistic effect for the relevant parties.

The aim of the SEA GHG Project meeting is to provide updates and feedback with SEA participating countries of their current status, gaps, challenges, barriers and capacity building needs (or technical assistance) in developing national GHG inventories for the third national communication (NC3) and first biennial update report (BUR1). There was also much feedback to WGIA.

2. Participation from USAID, USEPA and AusAID

The participants of WGIA are not only WGIA members—the US Environmental Protection Agency (USEPA), US Agency for International Development (USAID) and the Australian Agency for International Development (AusAID) have also attended. WGIA enables sharing of information on many donors' progress and the needs of the WGIA countries, which assists in coordination. The USA has conducted some projects, such as the SEA project, USAID Low Emissions Asian Development programme (LEAD programme) in Southeast Asia. Australia conducted a study tour for GHG improvement with Indonesia's government, which involved visiting facilities related to application of countermeasures for climate change through the WGIA network.

3. Mutual Learning Between Japan and Korea

The mutual learning between Japan and Korea is the first activity not involving WGIA and was held on the waste sector between GIO and Korea Environment Corporation (KECO) in the annual workshop in Korea in 2008. Korea's GHG inventory compiler invited Japan's counterpart to review its waste sector in terms of GHG inventory. Such mutual learning is a two-way process and does not involve one-way communication such as is found in the examiner–examinee relationship. As such, Japan checked Korean GHG inventories, but also Korea checked Japanese GHG inventories and gave Japan some comments. The comments from Korea contributed to improve the transparency of Japanese GHG inventories. The second mutual learning was held on the waste sector between Japan and Korea in Japan in 2009, and the third mutual learning was held on all sectors between Japan and Korea in Korea in 2010. Many findings resulted, which were not subjected to the

UNFCCC review, and thus also contributed to improved transparency of each other's GHG inventories.

The Secretariat of WGIA introduced this activity in WGIA8 in 2010. With the agreement of the participants, ML has been held in the WGIAs that followed as one of the sessions. For Non-Annex I Parties not mandated to be reviewed by UNFCCC, no particular attention needs to be paid to GHG inventories after submission. Previously, Non-Annex I Parties had never studied another's GHG inventories, which is where mutual learning provides an opportunity to study and learn from others' GHG inventories, which contributes to overall improvement of a country's own GHG inventories. Emission factors which other countries have developed and implemented to improve their GHG inventories, as well as issues concerning institutional arrangements which other countries face, and so on can be shared via ML. After ML, Non-Annex I Parties recognised the need both for the information in order to compile their own GHG inventories and the information on other countries, for comparison. Transparency and comparability are thus improved, and such findings lead to overall improvements in the GHG inventory.

4. Mutual Learning Plays a Role as External Quality Assurance (Korea and Lao PDR)

Korea, which is not included in Annex I Parties, does not have a responsibility to be reviewed by UNFCCC. And, as mutual learning does not employ any procedures such as UNFCCC reviews and only uses intercountry evaluations, it improves GHG inventories across the board. Korea implemented mutual learning as a form of external quality assurance in the WGIA sessions. Lao PDR, which attended the mutual learning sessions WGIA9 and WGIA11, also introduced mutual learning as a programme that emphasises peer reviews of the LULUCF with GIO. Lao PDR commented that this enhanced the accuracy and completion of the inventory of the LULUCF sector of Lao PDR.

5. Similarity Between Mutual Learning and International Consultation and Analysis (ICA) Procedure

Mutual learning involves "reading" a partner's GHG inventories in detail and studying other GHG inventories of other countries. As described above, mutual learning plays a role in the form of external quality assurance. In other respects, the ICA process of BUR is similar to quality assurance in that it is conducted by a third party, although it may not be regarded as quality assurance. ICA provides Non-Annex I Parties which lack sufficient human resources of quality assurance new opportunities to improve the quality of their GHG inventories

Mutual learning, just like ICA, contributes to improved transparency and comparability to evaluate the country-specific emission factors developed.

6. Cooperation with JICA Projects

GIO has collaborated with the Japan International Cooperation Agency (JICA) to build the capacity required to conduct periodical GHG inventories of developing countries. Projects have been implemented in Vietnam, Indonesia and Thailand to date. GIO provides leaning of technical issues of GHG inventories and has formed a relationship between JICA officers, GHG inventory compilers and expert WGIA participants. An author of this paper worked in a project in JICA Indonesia named Project of Capacity Development for Climate Change Strategies in Indonesia and lived in Indonesia for 2 years. The Ministry of Environment of Indonesia was well acquainted with GIO and respected GIO's experience and capacity. Making GHG inventories requires a great deal of networking and good connections, such as with ministries and researchers, as such can enable work to proceed smoothly.

In 2014, the Workshop on Capacity Development on Greenhouse Gas Inventory in the Southeast Asia Region entitled "How can CITC break through GHG inventory barriers?" was held as a back-to-back session of WGIA12, and GIO supported Climate Change International Technical and Training Center (CITC) and Thailand Greenhouse Gas Management Organization (a public organisation). This event represented the launch pad for CITC, and many participants of WGIA12 remained afterwards to attend this event. CITC is a training centre for other developing countries and was established by TGO as part of south–south cooperation.

11.2.4 Networks

1. South–South Cooperation: Thai–Myanmar Co-learning

A study visit of Myanmar's inventory compilers to Thailand for sharing information on measurement methodologies on the agriculture sector was held via the WGIA network. In the visit Myanmar learnt about measurement of GHG emissions from crop residue burning and rice straw burning from a Thai academic. Therefore, intercountry cooperation between neighbouring Non-Annex I Parties with similar socioeconomic or climatic conditions enhances regional cooperation and improves both parties' national GHG inventories.

2. Mutual Learning Between Australia and Indonesia

In 2012, Australia conducted a study tour for GHG improvement, which involved inviting Indonesia's government to observe a facility applying climate change countermeasures. In return, Indonesia's government invited Australia to check its GHG inventory in Indonesia. This is called mutual learning. The idea and importance of mutual learning to improve the quality of the GHG inventory originated at WGIA. Australia and Indonesia have communicated through the WGIA network.

3. Cooperation with Asia-Pacific Integrated Model (AIM)

GIO is part of the National Institute for Environmental Studies (NIES). A further Asian-related research team exists within NIES—the Asia-Pacific Integrated Model (AIM) team. The AIM team is involved with predictions of GHG emissions. The GHG inventory itself provides the key data needed for policy development to identify the major sectors, and the data in the inventory is also needed for prediction of GHG emissions and removals. The function of GHG emission predictions is one of the key benefits of the GHG inventory; thus, the AIM team is invited to the WGIA every year.

Many government officials who perform policy development attend WGIA as they need information on GHG emission predictions. Further, in developing countries, many GHG inventory compilers and experts are also in charge of policymaking and prediction of GHG emissions; thus, the one-workshop discussion covering GHG inventory and prediction is very useful for them.

Furthermore, participants from Myanmar requested GIO to introduce AIM team to them in order to learn more about prediction of GHG emissions. WGIA is thus the hub of Asia's climate change network.

11.2.5 Achievements

WGIA has produced several publications, as below:

- Promoted WGIA activities at a side event of SB24, COP15, COP19 and COP20
- Published a WGIA activity report "Greenhouse Gas Inventory Development in Asia – Experiences from Workshops on Greenhouse Gas Inventories in Asia"
  - (a summary report of 1st–4th WGIA)
- Proceedings of every WGIA

11.3 Other activities of WGIA

11.3.1 Website and Mailing List

In order to share and archive the information, GIO developed a website which is updated by workshops. PDF files of presentations given at the workshop and proceedings of WGIA can be downloaded for each year. These documents contain valuable and unique information and are thus downloaded all over the world, especially the information on least-developed countries. Documents for downloading can be found at the link below:

http://www-gio.nies.go.jp/wgia/wgiaindex-e.html

A mailing list of persons related to GHG inventory in Asian countries is prepared and managed by GIO. This mailing list is used to announce COP events and other international meetings. If participants wish to distribute information, such as conferences and events in COP, this can be done so via WGIA-Mailing List. Information exchanged through WGIA-Mailing List may be posted on the WGIA website for public access if so requested or considered useful. Also, the WGIA network platform can be used in parallel with other existing network platforms to complement them and should not be regarded as a replacement or competitor.

11.3.1.1 Brief Background Information on WGIA-Mailing List

The Greenhouse Gas Inventory Office of Japan (GIO) at the National Institute for Environmental Studies (NIES) developed the WGIA-Mailing List to serve as a primary support channel and to provide an opportunity for the WGIA community to share experience, knowledge and resources, voice concerns, seek advice and discuss topics of interest related to greenhouse gas inventories.

This is an initiative established for the WGIA community as an online regional network platform that all may take full advantage of through participation, according to the conclusions reached at Session III "Networking Experts in Region" at WGIA5. Currently, all participants in each WGIA subscribe to this mailing list.

11.3.2 WGIA-EFDB (Emission Factors Database)

In WGIA, participants share their experience with a focus on the estimation methods used in the GHG inventories, key category analysis and ways to address the problems faced in GHG inventory preparation and development to date.

In WGIA5, 2007, participants noted the utility of continuous and improved networking with stakeholders. Malaysia, Cambodia and Indonesia stated that a continuous database of emission factors for GHG inventory was needed. At that time no system for collecting GHG inventory data, such as activity data and emission factors, existed in developing countries. In Asia, some countries used emission factors of neighbouring countries with similar climate conditions, but there was no system or means by which to share such information. There was, therefore, a broad-based opinion concerning the need to develop a database for emission factors in order to share country-specific emission factors developed by WGIA participants.

Since WGIA7, GIO has collected a number of papers presenting country-specific emission factors developed for the various sectors. These values should be integrated in the Emission Factors Database (EFDB) being developed for the region.

One of the activities for the sectoral working group workshop during WGIA6 was to analyse data entered into the EFDB by closely scrutinising what environmental conditions, management practices or specific circumstances were developed for the EF. This meant that experts could discuss if an EF developed for another country may be "applicable" or "appropriate" for use in their own inventory. In this regard, experts could include, via adding remarks at the time of data entry in the EFDB, details as to what countries are "appropriate" for use of the EF factors, which would help other GHG inventory staff in the various countries in the region.

11.4 Conclusions

11.4.1 Importance of Ongoing, Face-to-Face Discussions

The Workshop on Greenhouse Gas Inventories in Asia (WGIA) has taken place annually since 2003 to provide an opportunity for countries in the Asian region to cooperate and share information and experience related to the development of national GHG inventories. WGIA is organised by GIO, a sustainable organisation. The 12th WGIA was held in Bangkok, Thailand, in 2014. Workshops were held in various host countries, as this leads to more attendees participating from Asian countries and enables more face-to-face discussion. The number of WGIA participants increased from 27 in 2003 to 130 in 2012. The latest WGIA (WGIA12) had an attendance of 123 from 14 WGIA member countries and international organisations in 2014. WGIA has become one of the biggest events for climate change in Asia. Through this ongoing face-to-face workshop, mutual trust among participants has been built.

11.4.2 Sharing of Information and Experience

In WGIA, the latest information and estimation methodologies are discussed, providing an opportunity for other countries to learn. Common problems and possible solutions are identified. This information helps participants compile transparent, accurate, time-series continuous, comparable and continuous GHG inventories. Many countries have overcome common problems and some countries have already developed country-specific emission factors, time-series GHG emission estimations and national systems as a result.

11.4.3 Network Utilisation

The WGIA network platform was established to exchange information on climate change and mitigation of GHG emissions as well as GHG inventory. WGIA's key function is to connect other activities among participants and Japan and also facilitate international cooperation; many collaborative activities, such as the SEA GHG Project, USAID, USEPA, AusAID as well as instances of mutual learning, have taken place as a result.

11.4.4 Continuity of WGIA

One of the reasons why WGIA can continue is its sustainability as an organisation (GIO/NIES). In addition, although the first workshop was just a small meeting, it has since grown, step by step, into something much larger. This environment of free discussion has built relationships of mutual trust among the participants. Holding WGIA every year has been enabled by the mutual trust built among and by the participating countries

References

Greenhouse Gas Inventory Office of Japan (GIO), CGER, NIES (2009a) Proceedings of the 6th Workshop on Greenhouse Gas Inventories in Asia (WGIA6) –"capacity building support for developing countries on GHG inventories and data collection (measurability, reportability and verifiability)" as a part of the "Kobe Initiative" of the G8 Environment Ministers Meeting, 16–18 July 2018, Tsukuba, Japan

Greenhouse Gas Inventory Office of Japan (GIO), CGER, NIES (2009b) Proceedings of the 7th Workshop on Greenhouse Gas Inventories in Asia (WGIA7) capacity building for measurability, reportability and verifiability under the Kobe Initiative, 7–10 July 2009, Seoul, Republic of Korea

Greenhouse Gas Inventory Office of Japan (GIO), CGER, NIES (2010) Proceedings of the 8th Workshop on Greenhouse Gas Inventories in Asia (WGIA8) – capacity building for measurability, reportability and verifiability- 13–16 July 2010, Vientiane, Lao People's Democratic Republic

Greenhouse Gas Inventory Office of Japan (GIO), CGER, NIES (2012) Proceedings of the 10th Workshop on Greenhouse Gas Inventories in Asia (WGIA10) – capacity building for measurability, reportability and verifiability-, 10–12 July 2012, Hanoi, Vietnam

Greenhouse Gas Inventory Office of Japan (GIO), CGER, NIES (2015) Proceedings of the 12th Workshop on Greenhouse Gas Inventories in Asia (WGIA12) – capacity building for measurability, reportability and verifiability-, 4–6 August 2014, Bangkok, Thailand

Ministry of the Environment, Japan and Greenhouse Gas Inventory Office of Japan (GIO), CGER, NIES (2014), National Inventory report of Japan, April 2014

UNFCCC, Decision 1/CP.16, The Cancun Agreements (2011)

Chapter 12
Japan's Comprehensive and Continual Support Package for the Creation of Scientific Climate Policies in Asia

Tomoko Ishikawa and Shuzo Nishioka

Abstract The response to climate change is a matter of increasing urgency, and from 2020, every nation will be required to reduce its GHGs. The unified reduction policies of the central governments of each country form the core of reduction policy implementation. Actual reductions are planned and implemented for each region and sector. As climate policies are strongly related to the development strategies and energy policies of each country, it is thus necessary for each country to independently mobilise knowledge to formulate strategies and policies based on domestic natural and developmental conditions.

The response to climate change has brought about a major turning point in modern civilisation, which was founded, and yet is still heavily dependent on fossil fuel energy. As Asian countries are currently in a period of strong growth, Asia as a whole must set a course towards low-carbon development that differs from the paths taken to date by developed industrialised countries. Science-based initiatives are indispensable to the formulation of climate policies, and in order for individual countries to frame policies and maintain ownership of them, scientific bases will, respectively, need to be created by each country.

From 2020, part of all-country participation in climate change mitigation entails INDCs (Intended Nationally Determined Contributions) be formulated. It is here that the achievements of a series of scientific cooperation projects promoted in the Asian region by the Government of Japan, in particular the Japanese Ministry of the Environment, can fully be appreciated.

Reducing GHG via scientific policymaking involves following the sequence of reduction target setting, reduction policy design, policy implementation, continuation and feedback (see Fig. 12.2). In order to carry this out, it is necessary to (1) ascertain GHG emission volumes for all processes (GHG inventories); (2) establish approaches to create unified climate policies for central and local governments (technologies, energy and GHG policy integrated assessment model, IAM); and (3) develop mechanisms to foster related research communities and strengthen

---

T. Ishikawa (✉) • S. Nishioka
Institute for Global Environmental Strategies (IGES), Kanagawa, Japan
e-mail: t-ishikawa@iges.or.jp

S. Nishioka (ed.), *Enabling Asia to Stabilise the Climate*,
DOI 10.1007/978-981-287-826-7_12

contributions therefrom to policy formulation (e.g. via strategic research programmes, fora for dialogue on policy and science).

In light of the growing importance of Asia in terms of global climate policy, the Government of Japan, together with other Asian countries, has promoted the creation of such scientific bases since the 1980s. These efforts have significantly assisted in policy formulation, including INDCs, in Asian nations. Further, the Low Carbon Asia Research Network (LoCARNet), comprising researchers directly engaged in climate policymaking processes in each country, was launched in 2012 in view of the rise in urgency of climate policy. LoCARNet has since organised relevant research communities based on ownership in each country to engage in the challenge of low-carbon development in Asia by facilitating knowledge sharing and cooperation throughout the Asian region.

Section 12.1 of this report describes the cooperation between Government of Japan and other Asian countries. Section 12.2 introduces in particular the activities of LoCARNet towards building research communities to promote concrete actions from 2020 as good practices to be disseminated throughout the world.

Keywords LoCARNet • Ministry of Environment of Japan • Scientific policymaking • Integrated assessment model • PDCA • Regional South-South-North Collaboration

Key Message to Policy Makers

- Asia holds the key to global climate stability.
- Science-based initiatives are indispensable to the formulation of climate policies.
- Government of Japan has promoted the creation of scientific bases in Asia since the 1980s, which has aided in formulating policy, including INDCs in Asian nations.
- LoCARNet has organised relevant research communities based on ownership in each country, to engage in the challenge of low-carbon development in Asia.

It is hoped Asia will take lead the way in a global transition to low-carbon societies, by establishing and implementing science-based policies.

12.1 Japan's Strategies to Support Scientific Climate Policymaking in Asia

12.1.1 Scope of Scientific Climate Policy

12.1.1.1 Scientific Context for Climate Policy

Based on observation results and model predictions, in the Fifth Assessment Report of the IPCC (AR5), Working Group I deemed that cumulative anthropogenic GHG emissions and global temperature increase have a proportional relationship (Fig. 12.1) (IPCC 2013: Summary for Policymakers. In Climate Change 2013: The Physical Science basis, p. 28). Carbon cycle research has shown that almost half of anthropogenic GHGs emitted are not absorbed and remain in the atmosphere. As the atmospheric lifetime of CO_2, which accounts for the majority of GHG, is thought to be more than 100 years, as long as emissions continue, the amount of CO_2 remaining in the atmosphere can only continue to rise. According to global warming theory, a rise in atmospheric concentration of GHGs directly results in a rise in temperature; therefore, as long as human-induced GHG emissions continue, so will the rise in global atmospheric temperature.

It is precisely because of the proportional relationship described above that we now face a critical issue—which is that whatever the temperature rise compared with the pre-industrial figure is, human-induced emissions must be brought to zero when such temperature is reached in order to stabilise climate. Ultimately, this means we must create a zero-emission world.

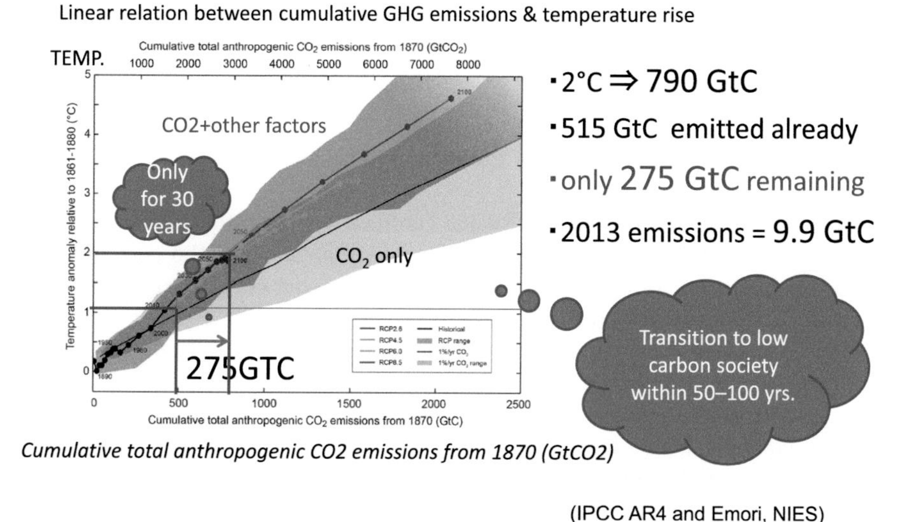

Fig. 12.1 Zero emission is the ultimate solution to stabilise climate

Agreements were reached at the G8 summit and UNFCCC COP16 (Cancun Agreements) of 2010 on a policy objective to limit the temperature rise to two degrees over pre-industrial levels, based on Article 2, 'Objective', of the Framework Convention on Climate Change, which calls for 'a level that would prevent dangerous anthropogenic interference with the climate system'. If the cumulative emissions corresponding to 2 °C are read from the IPCC/AR5 proportional graph, from which the cumulative anthropogenic emissions already released to date are subtracted, the amount of emissions permissible for a 2 °C increase is no more than around 30 years' worth of global emissions based on the emissions for 2010. Under these circumstances, the mission of the current generation should therefore be to be as frugal with this limited allowance as possible and, while evolving through the required stage of low-carbon society before this 'emissions budget' is used up (likely 50 to 100 years), also aim to create a zero-emission society for the whole world. The IPCC Working Group III has indicated the feasible emission pathway, namely, one that would reduce current global emissions (40 billion tonnes CO_2 equivalent) to half (20 billion) by 2050 (IPCC 2014: Summary for Policymakers. Working group III: Mitigation of Climate Change, p. 11).

If this 20 billion tonne allowance is distributed according to the projected population in 2050 of 10 billion, per capita CO_2 emissions are calculated to be about 2 tonnes. However, the reality is that per capita emissions have already topped 17 tonnes in the United States, 9 in Japan, 5.5 in China, 3 in Thailand, 1.6 in Indonesia and 1.4 in India. These figures reveal that almost all of these countries need to draw up policies to reduce GHG emissions. This represents a major transition challenge for developed countries, which were founded on, and at the same time are struggling to be free from lock-in of highly energy-consuming technologies, as they will need to overhaul their social infrastructure to one based on low-carbon society. Conversely, the major challenge for developing countries is their need to discover new, low-carbon development pathways that leapfrog over those utilised by developed countries to date.

12.1.1.2 Scope and Processes of Policy and Scientific Basis

What kinds of policies are needed when confronted with a major transition to a low-carbon world as described above? As energy policy is at the core of GHG emission reduction policy, it goes without saying that controls on energy consumption and a change in the structure of primary energy supply are required. However, policy cannot stop there—transitions are required in all sectors related to consumption and supply, including cities, land use, residential, transport and industry. The various sectors that must be covered by climate policy are indeed wide-ranging.

Formulation of long-term climate mitigation policies is carried out with the GHG emission reduction as an axis following the procedures shown in the middle of the figure below (Fig. 12.2): target setting, policy formulation, policy evaluation, monitoring of implementation results and feedback on the policy overall.

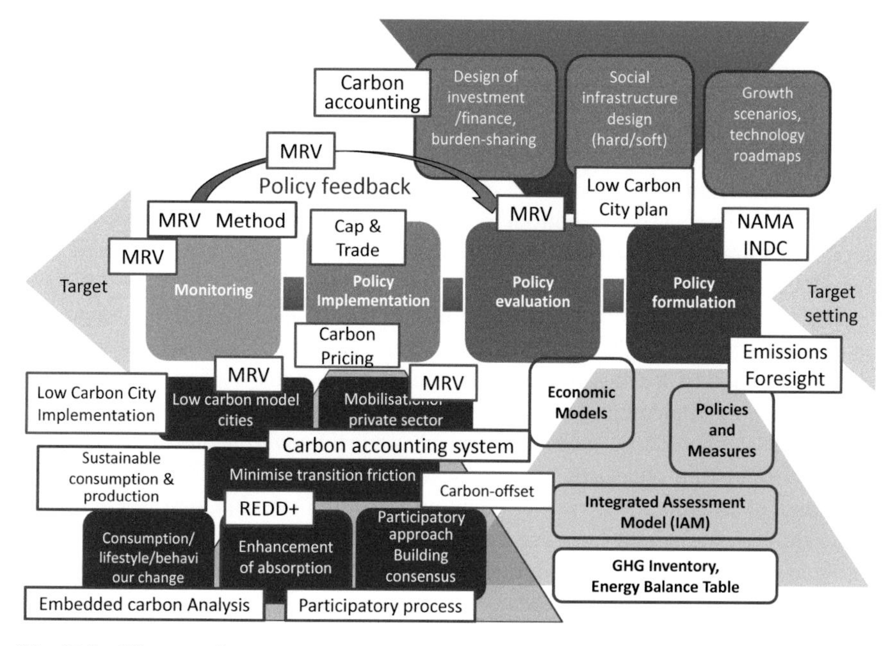

Fig. 12.2 Climate policy sequence and scientific support tools

Reduction targets are often decided a priori, as international agreements of the UNFCCC (e.g. reduction targets of the Kyoto Protocol) or as decisions by top management (e.g. the 26/41 % reductions of former President Yudhoyono of Indonesia). Leading up to these decisions are deliberations on approximate reduction outlooks and setting of rough targets based on such. When the world shifts its gaze in the direction of low-carbon society, as discussed above, the 2 tonne CO_2 per capita by 2050 figure will constitute a solid basis for reduction targets.

Subsequently, emission reduction scenarios based on reduction targets are needed. Based on economic growth rates and demographics, the necessary amount of services by sector for continuation of conventional policy (BaU: Business as Usual) and the energy needed to enable these services are calculated. These figures are then cross-referenced with the appropriate primary energy utilising an energy balance table, and then by multiplying GHG emissions per unit of primary energy, the GHG emissions for the entire nation can be calculated. Determining volumes of various activities mainly uses statistical data, and if the amount of energy per unit of activity and the amount of GHG generated per unit of energy are known, GHG emissions can be estimated. In sum, this series of estimations, termed 'inventory computation', is the bedrock for science-based climate policy.

These consolidated inventories as described above are made for each sector. In order to maintain consistency amongst sectors and create an integrated policy scenario, the Integrated Assessment Model (IAM) is indispensable. Included in this model group are the energy technology list and the GHG emission reduction cost curve covering all technologies, allowing for calculation of additional investment amounts for the entirety of reduction policies and overall costs.

A range of measures are available to bring the BaU emissions calculated in this matter closer to the reduction target amount, including regulatory methods such as establishing emission caps by sector and economic methods, including a carbon tax. With the application of such policy instruments, the system of reduction policy is determined (see Fig. 12.2, bottom right). The resulting investment cost in the necessary infrastructure based on the overall reduction plan can be estimated (see Fig. 12.2, top right).

These measures are then evaluated according to their impact on long-term economic growth via CGE (Computable General Equilibrium) modelling, and efforts are made to coordinate them with higher level plans.

Measures to implement policies are entrusted to the parties involved in actual reductions (stakeholders), such as municipal and local governments, industries and citizens. Reduction measures are advanced in respect of actual situations; local governments execute them via city planning and administration, and rural villages do so through forest and land-use plans. Likewise, reform in industrial structure, resource efficiency in manufacturing and distribution in the industrial sector and energy conservation measures for offices and households take place. Measures in civil society include rational consumption—consumption based on maximised utility or benefit of products or services.

The PDCA (plan, do, check and action) assessment cycle is applied to the whole process; results of actions by each stakeholder group are consolidated periodically to undergo MRV (measurement, reporting and verification), whereby feedback is given on reinforcement of measures and changes to plans. Through PDCA, inventories and integrated assessment models used in the early stages of planning are effectively utilised as criteria to determine efficacy.

The above illustrates how substantial the scientific base is required to be for GHG reduction policy development. This base includes GHG inventories, policy formation based on integrated assessment models, economic assessment methods for policy, knowledge on policy formation and infrastructure building at the city level, calculation of resource efficiency by Life Cycle Assessment (LCA) methods and analysis of public behaviour. Further, as geographical, economic, resource and political factors vary across the region, so do the required scientific bases; therefore, each and every country needs to create domestic climate policies by fostering domestic research communities in order to realise scientific bases in accordance with domestic environments.

According to IPCC AR5, climate change is already progressing, and its impacts are evident around the world. Some countries have already been affected and have initiated adaptation activities. This makes the need to share scientific knowledge all the more important, not only in terms of mitigation but also adaptation.

12.1.2 *Japan's Support for Climate Policies in Asia*

12.1.2.1 Japan's Policy on Asian Cooperation

In the Business as Usual (BaU) case, Asia is expected to account for half the world's economy, energy consumption and GHG emissions by the year 2050, thus the region holds the key to global climate policy in both aspects of GHG emission reduction and adaptation.

Japan, with its strong ties to neighbouring countries in Asia, both geographically, culturally, historically and economically, has since the 1980s actively and continuously carried out initiatives in knowledge sharing to form scientific bases for climate change policy formulation. These efforts have already born fruit in Asia's developing countries and in the national communications and Biennial Update Reports (BUR) submitted to the UNFCCC. A strong foundation has also been laid for preparation of Nationally Appropriate Mitigation Actions (NAMA) and Intended Nationally Determined Contributions (INDCs), which are to start in 2015.

Japan's stance on climate change can be summed up in the following five principles: (1) adherence to international polluter pays principle (PPP), upholding Japan's responsibility; (2) active contribution as Environmental Nation, contributions that capitalise on Japan's capabilities in environmental protection; (3) full respect for partner's autonomy, support for self-reliance and autonomous environmental management of other countries; (4) contribution to global environmental diplomacy, exercising leadership in the environmental arena; and (5) positive facilitation of international agreements, preparing frameworks for international policy agreements.

The above five principles for action were discussed in the COSMO Plan (Nishioka, S., 1990: Policy scientific insight required for responding to Global Warming, Environmental Research Quarterly No. 77, January 1990, 14–20 (in Japanese)) in early 1990 (Comprehensive Strategies for Moderating Global Warming plan). Likewise, cooperation with other Asian nations on climate policy and support for policy formulation based on individual respective country ownership are also part of Japan's basic stance.

Japan's cooperation is based on forward thinking and long-term capacity building through sharing of scientific knowledge and world affairs, preparation of inventories, support for policy formulation and human resource development in science and policy. To such end, Japan believes countries in Asia should not rely on foreign consulting bodies and instead take ownership of policy formulation, based on the conviction that this will lead to enduring low-carbon development in Asia. This support from Japan has been acknowledged by Asian nations to date and has bestowed bonds of trust.

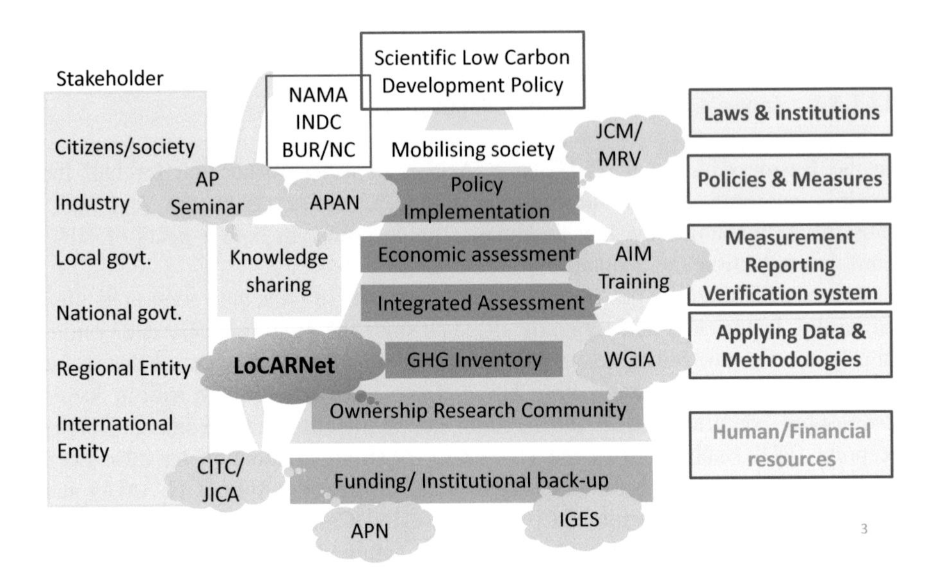

Fig. 12.3 Elements supporting scientific low-carbon development policy and Japanese collaboration with Asian countries

12.1.2.2 Support for Comprehensive, Continual and Systematic Creation of Science-Based Policies

The Ministry of the Environment of Japan has maintained, without interruption, its support for systematic science-based policy formulation in the Asian region, embracing all fields related to climate change (Fig. 12.3) (Table 12.1).

Climate change first caught the attention of science in the late 1970s, and in 1988, the IPCC was formed to evaluate its scientific aspects. In 1989, the Government of Japan held the 'Tokyo Conference on the Global Environment', which brought together experts from around the world to deliberate on how Japan would contribute to preserving the global environment and, in particular, how it could contribute to the climate change issue. While this event was over two decades ago, strong growth in Asia had already been anticipated at the time. Japan, with its strong ties in Asia, thus set out to form a system of technological and scientific cooperation based on the symbiotic viewpoint of mutual benefit.

Shared Recognition of Global Trends in Science and Policy Based on the fact that climate change is caused by human interference with nature, responses must begin by firstly deciphering what is happening in the natural environment and then sharing this insight—scientific knowledge. This is just what Japan did, as immediately after the release of the first IPCC report in 1990, GoJ launched the 'Asia-Pacific Seminar on Climate Change' for knowledge-sharing amongst Asian nations. The meeting, which convened environment policymakers from several Asian countries, was jointly sponsored by UNEP/UNCRD and included Chairman Bolin of the IPCC, as well as Chairman Izrael and Vice-Chairman Hashimoto of Working

Table 12.1 Japan's comprehensive, continuous and systematic support for science-based climate policy development in Asia

Objectives	Activities	Contents	Est. since		Outputs
Knowledge sharing	Tokyo Conference on the Global Environment and Human Response toward Sustainable Development (UNEP and Japanese Govt.)	Advisory meeting on global climate change	1989		
	Asia-Pacific Seminar on Climate Change (AP Seminar)	Knowledge sharing on IPCC results and UNFCCC progress	1991–	23rd meeting in 2014	Approx. 800 participants in total
		Support for developing plans addressing CC		Rotating presidency	
Formulation of basic grounds in research	The Asia-Pacific Network for Global Change Research (APN)	Promoting joint research and capacity development in Asia	1996–	Funds: 400 million yen (includes 120 million yen for new CAF projects in 2014)	Approx. 225 projects
		Low-carbon initiatives			
Common research funds		Climate adaptation framework			
	Institute for Global Environmental Strategies (IGES)	(Approx. 100 researchers)	1998	Constant research	White Paper, ISAP (annual conference).
		IPCC/NGGI/TSU	1999	Global standard	
	Activities as Regional/International Environment Research organisation at the initiative of PM in Japan				Many joint researches in Asia
Capacity development in forming the basis of climate policy	Workshop on Greenhouse Gas Inventories in Asia (WGIA)	Improving accuracy of GHG inventories in Asia	2003–	Annual meeting	Approx. 900 participants in total
		Sharing knowledge and experience amongst researchers and policymakers		12 times in total	
		Promoting regional cooperation			

(continued)

Table 12.1 (continued)

Objectives	Activities	Contents	Est. since		Outputs
	Asian Pacific Integrated Model (AIM) Int'l WS/Training workshop	Training on integrated models for developing climate mitigation policies	1995–	19 times (annually)/irregular, more than ten times	Approx. 500+ participants in total/ Approx. 100+ in total
		Organised by NIES and Kyoto Univ.	1997–		
	Asia Pacific Adaptation Network (APAN)	Asia-Pacific Climate Change Adaptation Forum	2009–	Biennial meeting	Approx. 1850 participants (First–third forum)
		Subregional/thematic meetings by APAN nodes			
	Low Carbon Asia Research Network (LoCARNet)	Formulating research community, policy dialogues between researchers and policymakers on climate policy, knowledge sharing	2012–	Third Annual Meeting in 2014	Approx. 250 participants
				20+ WSs	~800 part
Climate policy and technical training	Support for the Climate Change International Technical and Training Centre	Training on climate policies for policymakers in central/local govt., practitioners, businesses mainly in ASEAN countries	2014–	Starting from Oct. 2014	
	(CITC Thailand)	TGO established CITC by receiving support from JICA			

Group II. Following UNFCCC establishment, this seminar has focused not only on science and continues to function as a forum for knowledge exchange amongst Asia's climate change policymakers and related international organisations on responses to the UNFCCC. Meetings rotate within Asian nations and the 23rd was held in 2014.

A Permanent Research-Dedicated Institution to Lead Environmental Research in Asia Under a prime ministerial initiative after the 1992 Earth Summit in Rio de Janeiro, the Institute for Global Environmental Strategies (IGES) was established in 1998. IGES is an institute dedicated to conducting research on environmental issues in Asia. With a staff of about 200, including researchers from both within and outside Asia and management, it operates to maintain the various cooperative research networks and implement research activities that are mainly focused on Asia.

Sharing Scientific Bases for Policymaking As shown in Fig. 12.2, a substantial amount of scientific data and assessment methodologies are required for policymaking in response to climate change. Of such, accurate ascertainment of GHG emissions and inventory work are particularly important. In recognition of this—and of the importance of setting fundamental processes at an early stage—GoJ, upon the request of the IPCC, established the Technical Support Unit (TSU) of the IPCC Task Force on National Greenhouse Gas Inventories at IGES in 1999. The IGES research units, in cooperation with the IPCC TSU and researchers from other Asian nations, led the task of amassing emission factors for Asia's unique forest soil and rice fields. In 2002, the Greenhouse Gas Inventory Office of Japan (GIO) was established in the National Institute for Environmental Studies (NIES). Using this office as a base, GoJ, in cooperation with IPCC TSU and the UNFCCC, launched the 'Workshop on Greenhouse Gas Inventories in Asia (WGIA)' in order to build capacity with the goal of ascertaining of GHG emissions throughout Asia. This workshop, which is hosted in a round-robin fashion by Asian countries, pairs up policymakers and researchers to form research communities that can maintain scientific accuracy over the long term. The latest data shared at the workshops is also reflected in national communications and policy formation in each country. The year 2014 marked the 13th year of this workshop, and since 2013, when BUR became obligatory, it has attracted over 120 delegates. These workshops have greatly contributed to estimating GHG emissions in Asia based on sharing of QA/QC methods and mutual learning.

Consistent Support for Policy Formulation Based on Integrated Assessment Models The application of integrated assessment models is indispensable in refining and coalescing the ideas of disparate ministries and agencies into consistent plans and policies for individual countries, including NAMAs and INDCs. The National Institute for Environmental Studies and Kyoto University began developing the Asia-Pacific Integrated Model (AIM) for climate policy in 1990, the goal of which is to promote its use in Asian countries, and joint research with China and India has taken place. Based on five years of research from 2004, this model is at present used in the drafting of Japan's plan for a low-carbon society. With the target of applying the model to ASEAN countries via 'Low Carbon Asia Research' from 2009, annual AIM Symposiums have been held jointly with researchers from each country to promote support for climate policy formation in each country. The model was applied to the creation of NAMA and INDC in Thailand and to the low-carbon city plan for Iskandar in Malaysia. In parallel with these activities, since 1994, 12 AIM training workshops (about 50 participants each) have been held, which are targeted at researchers and policymakers in Asia to foster development in climate policy at both the national and city levels.

Formation of a Research Community to Support Low-Carbon Development Policy in Each Country The Low Carbon Asia Research Network (LoCARNet: see Sect. 12.2 of this paper) forms a community for research on low-carbon societies with ownership by each respective country and aims to directly support policy in each country and to promote science-based low-carbon development policy throughout the region via mutual cooperation in the Asian region and South-South cooperation. The network was proposed at the ASEAN + 3 Environmental Ministers Meeting and has been active since 2012. Considering the urgent nature of climate policy, the network is led by researchers already deeply involved in policy support. Policy dialogue workshops between policymakers and research communities have been held in various countries. At the network's annual meetings, which began in 2012, discussions take place on key research topics for low-carbon development. At the annual meeting in Bogor in 2014, in the Bogor Declaration entitled 'Asia Is Ready to Stabilise Climate', researchers highlighted efforts towards climate stabilisation in Asia and also reported on the potential for reductions in Asia and good practices in Asia to provide input to international policymaking processes such as the UNFCCC. In order to become an independent leading network for low-carbon policy research in Asia, the formation of a CoE (centre of excellence) alliance is underway. The network aims to foster research communities in countries that currently lack them—Cambodia, Lao PDR and Myanmar—through South-South cooperation.

Funding for Climate Change Policy Research The Asia-Pacific Network for Global Change Research (APN) is a research fund that promotes not only research on climate change but also on the overall global environment, including biodiversity, transboundary air pollution and marine pollution. Activities were launched in 1996 on the initiative of Japan and with funds provided by the United States, Australia, New Zealand and the Republic of Korea. APN provides funding for joint research amongst Asian researchers. From 2013, the Low Carbon Initiatives (LCI) fund was established to accelerate low-carbon research in this region.

Knowledge Sharing on Climate Change Adaptation Upon the launch of the IPCC, Japan served as the Vice-Chairman of Working Group II (impact assessment) and was in charge of the chapter 'Technical Guidelines for Addressing Climate Change Impacts and Adaptations' of the Second Assessment Report in 1995. Utilising the Guidelines released in 1994, research on impacts and adaption was carried out in many developing countries. Japan also took the lead in the chapter on impacts and adaptation in Asia for the Third and Fourth reports by forming a research community for the same in the Asian region based on APN research funds.

As the impacts of climate change vary from place to place, adaptation measures suited to each respective setting are needed, and such measures can be bolstered through shared experience. This was the concept behind the Asia Pacific Adaptation Network (APAN), which was proposed by the Ministry of the Environment of Japan and established in 2009 by a Thai prime ministerial declaration. This network is a forum for knowledge sharing between researchers, policymakers and experts engaged in the field of adaptation measures in Asian nations, and its activities are overseen by a secretariat in Bangkok. The influence of APAN as a role model of good practice has extended to other regions of the world—the creation in 2013 of the Global Adaptation Network (GAN) by United Nations Environment Programme (UNEP) is one example.

Developing Human Resources for Low-Carbon Policy Formation Since the 1980s, the Ministry of the Environment of Japan, in cooperation with the Japan International Cooperation Agency (JICA), has focused its efforts on human resource development to support environmental research and policy in Asia. Results of such are the Environment Research and Training Centre (ERTC) created in Thailand in 1989, the Environmental Management Centre (EMC) in Indonesia in 1990 and the Sino-Japan Friendship Centre for Environmental Protection in 1990. Recognising the importance of Asia's response to climate change, JICA provided assistance for establishing the Climate Change International Technical and Training Centre (CITC) under Thailand's Ministry of Natural Resources and Environment (MONRE) and the Thailand Greenhouse Gas Management Organisation (TGO). This centre, which began operating in May 2014, conducts training for policymakers, experts and industry representatives from the ASEAN in the areas of GHG inventories, low-carbon societies, mitigation measures and technology and adaptation measures. Leading researchers from the ASEAN region of LoCARNet have advised on training content and participated in curriculum development and lectures. Both IGES and the LoCARNet have participated as JICA experts.

12.2 The Low Carbon Asia Research Network (LoCARNet)

12.2.1 Science-Based Policy Formulation: LoCARNet Research and Policy Integration Activities

In keeping with trends in international discussions, the countries of Asia continue to make steady progress in developing low-carbon plans and strategies based on a green economy. The Low Carbon Asia Research Network (LoCARNet; secretariat:

IGES) supports development planning and strategy building by researchers and research institutes in various Asian countries in cooperation with the National Institute for Environmental Studies and Kyoto University. By setting up opportunities for researchers and policymakers to engage in discussion, each country can show how it is quantitatively reducing GHG and take planned actions to enable more effective policymaking for low-carbon development. Further, because countries in the region share a common economic footing and geographic location, researchers in each country can pool research results, carry out knowledge sharing and actively promote a system of mutual learning to facilitate South-South cooperation. Below is an overview of LoCARNet.

12.2.1.1 Background, Sequence of Events, Organisation and Policies

Background and Policies

LoCARNet is an open network of researchers, research organisations and like-minded relevant stakeholders that facilitates the formulation and implementation of science-based policies for low-carbon development in Asia.

A new international framework currently being considered by UNFCCC includes all GHG emitter countries from 2020 and beyond. In order to stabilise global climate, in addition to developed countries being required to drastically reduce their GHG emissions, those of developing countries will also need to be reduced based on their predicted increases. If trends in mass energy consumption and growth continue, by 2050 Asia will account for half of total global emissions, which will also cause heavy concomitant impact on the region's populations. On the other hand, if the substantial investments are redirected towards creating low-carbon societies, Asia could lead the world in low-carbon development. Right now, we are standing at the crossroads.

All countries in the Asian region are currently working on plans to achieve low-carbon development. Numerous policy steps are involved, and the formulation of such plans requires scientific knowledge spanning a broad spectrum of fields. It is thus plainly apparent that an interdisciplinary community for researchers and research organisations needs to be formed. Also, as policies for low-carbon development involve important decisions that determine a country's future potential, each country should have the right to self-determination in formulating these policies, therefore must have its own robust, scientific research footings.

While the above calls for establishing unique foundations, commonalities exist in the regional environment and in stages of development. In this regard, the exchange of scientific knowledge amongst researchers in Asian countries in respective areas of expertise will greatly promote science-based low-carbon policies, and reinforcing the scientific base to bolster the formulation of policies for low-carbon development in each country in the region will require a quantum leap in cooperation, not only from researchers and research communities but also from international organisations, donor agencies, NGOs and other like-minded relevant stakeholders. The ultimate aim of LoCARNet is to promote regional cooperation to facilitate the formulation and implementation of science-based policies for low-carbon growth in the Asian region, together with relevant stakeholders.

LoCARNet effectively promotes research on low-carbon growth policy by enabling effective dialogue between scientists and policymakers and also encourages domestic collaboration amongst researchers whose research capacities and scientific knowledge are firmly grounded in their home countries (ownership in country by these researchers). LoCARNet also aims to increase research capacity in the region through knowledge sharing and information exchange, in the context of not only North-South cooperation, but also South-South regional cooperation (Box 12.1).

Box 12.1 Unique Characteristics of LoCARNet

Unique characteristics of LoCARNet

- LoCARNet is a network of leading researchers, research organisations and like-minded relevant stakeholders deeply involved in low-carbon growth policy processes in Asia.
- Science-Science-Policy Dialogue: LoCARNet promotes research on policies for low-carbon growth by enabling sufficient dialogue between scientists and policymakers.
- Country-based ownership of knowledge: LoCARNet encourages collaboration between researchers in-country whose research capacity and scientific knowledge are firmly grounded on home soil.
- Regional South-South–North Collaboration: LoCARNet aims to increase research capacity in the AP region through knowledge sharing and information exchange as a part of regional cooperation—not only North-South but also South-South cooperation.

Establishment of LoCARNet

From 2009, Japan's Ministry of the Environment and National Institute for Environmental Studies embarked on a research programme on low-carbon development in Asia. Under this coalition, IGES, together with NIES and Kyoto University, has been conducting workshops that promote dialogue between policymakers and researchers in Indonesia, Thailand, Cambodia and Malaysia, as well as networking amongst researchers in the region to encourage low-carbon development in Asia. During the course of these workshops, it became clear that low-carbon development in Asia must take place.

Considering the significance of Asia for global climate policy, the Government of Japan and IGES proposed the creation of LoCARNet at the October 2011 ASEAN + 3 Environmental Ministers' Meeting in Cambodia. The launch took place in April of the following year at a side event ('East Asia Low Carbon Development Knowledge Partnership', organised by three institutes—National Institute for Environmental Studies, Japan International Cooperation Agency (JICA) and IGES) one day prior to Japan's Ministry of Foreign Affairs' 'East Asia Low Carbon Growth Partnership Dialogue'. It was then officially reported to the Dialogue the following day. As IGES had been commissioned to organise the Secretariat of the International Research Network for Low Carbon Societies (LCS-RNet) based on a decision made at the G8 summit in 2008, it was also tentatively placed in charge of the Secretariat function of LoCARNet as well.

Recent trends point to initiatives in low-carbon, green growth taking place in many parts of the region, and such are supported by developed countries and international organisations. Concurrently, LoCARNet will facilitate the creation and accumulation of knowledge to help formulate and implement science-based policies for low-carbon growth in the Asia region.

Scope of Activities

The knowledge required and the issues related to promoting low-carbon growth policies are extremely diverse in nature. As such, the network needs to focus its sphere of activities on priority areas in order to provide efficient and effective results.

Regionally: ASEAN Core, Centred on Asia, Futuristically Global Although the significance of Asia was previously mentioned, China has already established a strong low-carbon policymaking process for itself and also leads the world in utilisation of renewable energies, both in terms of facilities and production. As regards India, its per capita GHG emissions are still low although it has the potential to become an emission giant in the future. Therefore, for the time being, the network will primarily focus on the parts of the ASEAN region that are undergoing striking development, with initiatives in this region making up its core activities. The network will, however, continue to promote research exchange that includes China and India.

Since the Copenhagen Accord, cooperation between developed and developing countries has intensified. In response to this, comparatively recently, momentum has been building for knowledge sharing around the world: initiatives similar to LoCARNet have been initiated by Europe, the United States and international organisations. In addition, LCS-RNet, promoted by Japan with the G8 countries, intends to expand its network to include emerging economies and developing countries, where GHG emissions are predicted to increase greatly in the future, while collaborating with LoCARNet in Asia. Further, as LCS-RNet has been requested to deal with integration of mitigation and adaptation, it is thought that

LoCARNet will assist in maximising collaboration in these initiatives to yield mutually beneficial and effective results.

Targets: Low-Carbon Research as the Core, Stronger Links with Policy and Industry The objective of the network is to further low-carbon development. To do so, it is necessary to promote scientific policy based on research. Hence, it is essential to first have a robust research community in each country. However, these communities are meaningful only when research is reflected in policy and industrial activities, making strong cooperation with government (both central and local government) and business sectors a must.

Stakeholders: Researchers at the Core, Cooperation with Policymakers and Participation of Supporters, Expansion of Related Parties The roots of the network will be grounded in its research capacity. The skills of researchers involved in low-carbon development must be consolidated and the research community expanded; as such, researchers will play the leading role. Deliberation on issues to be addressed must be made from the viewpoint not only of scholarly but also policy aspects. Also, support from policymakers and the role of funding agencies are substantial in terms of finance and organisational aspects. In order to disseminate outcomes and bring about impacts, the cooperation of a broad range of stakeholders, including international organisations and NGOs, is essential.

12.2.1.2 Activities and Outcomes

The activities of LoCARNet can be categorised under three pillars. Here again, the countries of Asia continue to make steady progress in developing low-carbon plans and strategies based on a green economy.

As the first pillar, LoCARNet has maintained policy dialogue between researchers and policymakers in selected countries, together with the National Institute for Environmental Studies and Kyoto University. For example, LoCARNet experts in Indonesia, working with the National Development Planning Board (BAPPENAS), have conducted an economic evaluation comparison of a low-carbon development draft plan; a LoCARNet expert in Thailand has worked closely with the Thailand Greenhouse Gas Management Organization (TGO) and reflected his analysis within Thailand NAMA development, and in Iskandar, Malaysia, a LoCARNet expert has been conducting low-carbon city planning and implementation in order to promote collaboration with local universities and implementing organisations.

Through these policy dialogue sessions, policymakers have gradually recognised the importance of in-country involvement of researchers and research communities in the policymaking process, and as a consequence, scientific policymaking has been promoted in these countries.

In parallel, as a result of bolstered human resources in Japan (National Institute for Environmental Studies (NIES), Kyoto University and IGES) engaged in supporting low-carbon development in developing countries, a series of steps has been established for low-carbon development planning, including development of the GHG inventory, vision development, quantitative scenario creation, economic evaluation, action plan design and road map formulation.

As the second pillar, having conducted several workshops for capacity development in Cambodia, LoCARNet organised a trilateral workshop for Cambodia, Lao PDR and Myanmar in February 2014 in Phnom Penh, Cambodia. This workshop aimed to have Cambodia, Lao PDR and Myanmar each utilising its own capacity to present quantitative GHG emission reduction potentials, to advance the organisation of research communities in-country, to enhance policymaking in a more effective manner and to provide a forum for researchers and policymakers to engage in discussion. In addition, owing to the various commonalities shared by the participating countries, including level of economic development and geographical characteristics, this workshop also provided opportunities for researchers in each country to bring together their research results, carry out knowledge sharing and actively promote a system of mutual learning that facilitates South-South cooperation. Through such activities, countries such as Cambodia, Lao PDR and Myanmar are projected to be in a better position to implement the low-carbon plans discussed in the workshop.

As the third pillar, LoCARNet organised a series of annual meetings—the first being in October 2012 in Bangkok (Thailand), second in July 2013 in Yokohama (Japan) and third in October 2014 in Bogor (Indonesia). These meetings underscored the importance of networks as fora for sharing knowledge in order to bring about low-carbon societies and low-carbon development. Further, a number of urgent issues for research common to the Asian region were discussed, including 'the need for capacity development towards the framework for 2020', 'comparison of reduction potential of Asian countries towards achieving the two degree target', 'the role of cities as pioneers for LCS', 'low-carbon technologies required in Asia', 'Asian issues: emission reduction in the agriculture, forestry and land-use sectors' and 'integration of low-carbon issues and climate change adaptation'. It is likely that LoCARNet will transfer these outputs from the research community to the policy training centres planned for ASEAN countries (Thailand and Indonesia).

In addition, knowledge sharing at the annual meetings and South-South regional cooperation together increase momentum towards realising low-carbon development in Asia, which could realise a significant GHG reduction potential and sends a very positive message. In November 2014, LoCARNet organised its Third Annual Meeting, held in Bogor, Indonesia, where researchers in Asia issued the 'LoCARNet Bogor Declaration', which states 'Asia is ready to stabilise climate' (Box 12.2).

Box 12.2 LoCARNet Bogor Declaration (November 2014)

LoCARNet Bogor Declaration

Asia Is Ready to Stabilise Climate

Recognising the huge risk of climate change to human well-being as predicted by science communities; welcoming the start of a new global regime to avert it; confirming the growing importance of regional cooperation for low-carbon transition; and drawing upon Asian wisdom to contribute in stabilising climate, the participants of the LoCARNet Third Annual Meeting reaffirm that:

1. Asia has research capacity; Asia has research networks that support policymaking.
2. Asia has the potential for low-carbon transition which is adequate to contribute to the two degrees temperature stabilisation target.
3. Asia has the technological, financial and institutional capacity to facilitate low-carbon actions.
4. Many 'good practice' examples exist and their replication is challenging. Continued technological and institutional innovations are needed to support the transition to a sustainable low-carbon society.
5. Asia is ready to make due contribution to global climate stabilisation.

However:

1. Diversity amongst Asian nations poses challenges for framing uniform policies, but provides opportunities for discovering a range of options. Regional cooperation for low-carbon research is therefore challenging as well as rewarding.
2. Asia houses a sizable fraction of low-income families. Their development needs require special attention to ensure that their welfare is not compromised.

Low-carbon research in Asia shows that timing is critical; lock-ins must be avoided, and all 'leapfrogging opportunities' should be seized and realised by positive actions, supported by global climate policies, including technology transfer and incremental finance.

Asia is ready for low-carbon transition and awaits signals from the Paris Climate Change Agreement to deploy actions towards climate stabilisation.

26 November 2014
Low Carbon Asia Research Network (LoCARNet)

12.2.2 Future Plan

Three years have passed since LoCARNet's official launch. During this time, research institutes and organisations in Asia have exhibited considerable development, and based on cooperation activities conducted so far, now is time for Asia to mobilise the necessary knowledge through its own capacity and to establish and manage an autonomous, regionally owned network.

Under the new post-2015 framework, respective governments in Asia should promptly develop long-term, low-carbon development plans and put them in effect. In order for Asia to develop its own policies with full ownership, the region must possess an autonomous regional research community. And, at the policy implementation stage, it is necessary for science (research) to be reflected in policies. Moreover, knowledge held in the scientific community must be leveraged to make relevant stakeholders, especially in business, and move towards low-carbon development. Under the LoCARNet scheme, several research institutes and organisations exist in this region, which are already working closely with policymakers and involved in the policymaking process in their respective specialised fields as centres of excellence (CoE). However, in the future, synergy effects amongst these institutes, as a CoE coalition or CoE alliance, must be maximised to drive progress in policies and in the business sector in this region.

In this transitional stage that includes 2015, it will be imperative to make policy proposals, in parallel, to each government, to the Asian region, to the global international community and to relevant stakeholders. This project will take the form of action research that develops the organisation through these kinds of on-the-job activities.

As regards future network activities, LoCARNet will focus on promoting priority research in fields common to the Asian region; facilitating financial and institutional support for fostering and strengthening research capacity in the region, in collaboration with donors, other institutions and organisations concerned; securing routes to reflect research results in policymaking; and activating networking and information dissemination with and to other like-minded stakeholders towards realising low-carbon societies and low-carbon development.

References

IPCC (2013) Summary for policymakers. In climate change 2013: the physical science basis, p 28

IPCC (2014) Summary for policymakers. Working group III: Mitigation of climate change, p 11

Nishioka S (1990) Policy scientific insight required for responding to global warming. Environ Res Q 77:14–20 (in Japanese)

Printed in the United States
By Bookmasters